An Illustrated Atlas of the Skeletal Muscles

· *Study Guide and Workbook* ·

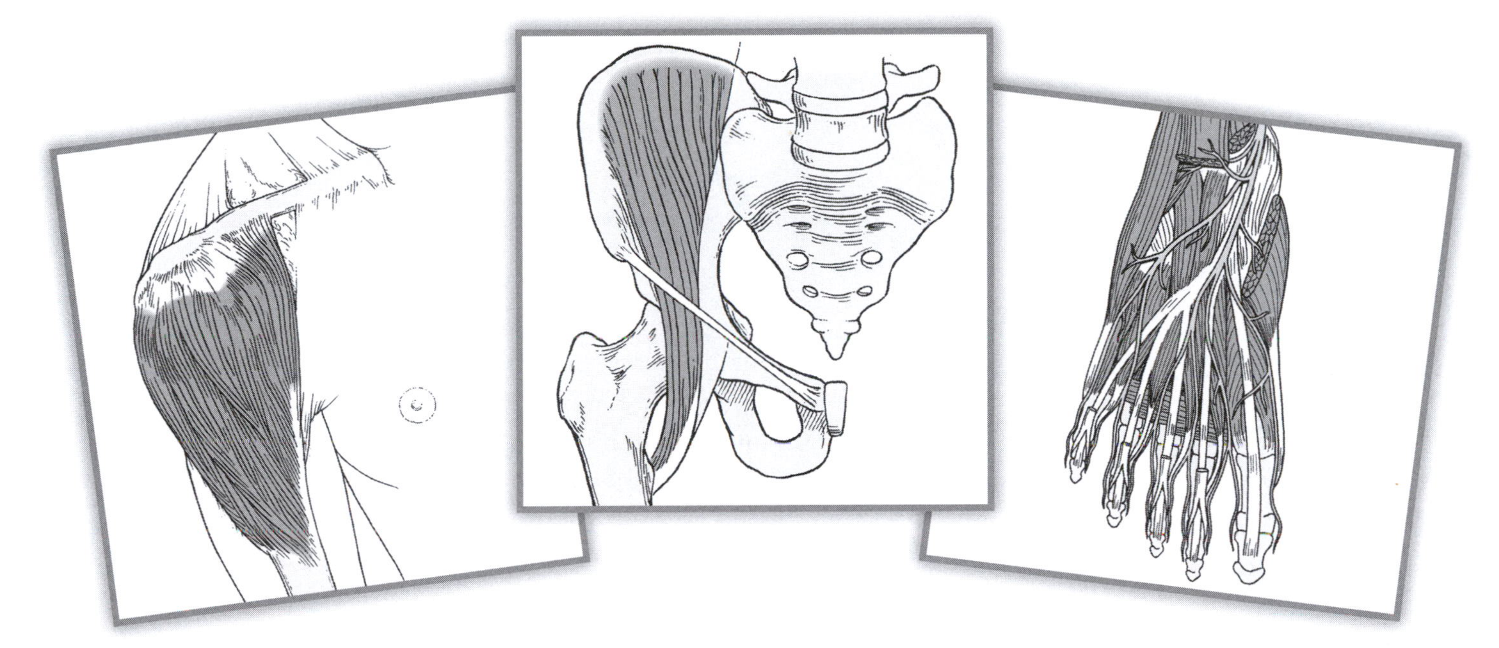

Bradley S. Bowden
Alfred University

Joan M. Bowden
Alfred University

Surgical Illustrator
Peggy Firth

Digital Graphic Artist
Michael North

925 W. Kenyon Avenue, Unit 12
Englewood, CO 80110
800-348-3777
www.morton-pub.com

Book Team

Publisher	:	Douglas Morton
Editor/Project Manager	:	Dona Mendoza
Interior Design/Production	:	Joanne Saliger
Copy Editor	:	Elizabeth Budd
Cover Design	:	Bob Schram, Bookends

Morton Publishing Company does not assume responsibility for any inaccuracies or omissions or for consequences from the use of information found in this *Workbook*. The materials provided are for information purposes and are not intended for use as diagnosis or treatment of a health problem or as a substitute for consulting a licensed or certified professional.

Library of Congress Control Number: 2011931624

ISBN: 978-089582-884-2

Printed in the United States of America

10 9 8 7 6 5 4 3 2 1

Preface

This skeletal muscle ***Study Guide and Workbook*** provides a significant review and reinforcement tool to aid students in mastering their knowledge of the human skeleton, articulations, body motions, and the innervations and actions of individual and functional groups of muscles. Although developed to accompany ***An Illustrated Atlas of the Skeletal Muscles***, this ***Study Guide and Workbook*** provides the same opportunities for all health professions and life science students looking for an effective self-study guide on these topics.

The variety of objective questions (illustration labeling, matching, sentence completion, multiple choice, fill in the blanks, short answers, and the hands-on palpation and case study exercises) engage students in looking at the structural components of the skeletomuscular system and their innervations from a variety of perspectives. This strengthens their understanding of how the components of the skeletal system interact and, together with the associated muscles, produce body movements in response to innervation. The diversity of question types and practical exercises also respond to different learning styles. Questions drawn from the Information Boxes in the ***An Illustrated Atlas of the Skeletal Muscles*** test student knowledge of relevant clinical issues, such as injection sites, injury and disease, and developmental anomalies.

Organization of the Study Guide and Workbook

This skeletal muscle ***Study Guide and Workbook*** follows the same sequence of chapters as in ***An Illustrated Atlas of the Skeletal Muscles***. Questions and exercises are organized in each chapter to move from the basic information progressively toward more complex information, enabling students to review individual facts that are integrated in later questions.

Muscle illustrations in Chapters 3–5, with muscle names, actions, points of attachment, and innervations on the reverse side, can be cut and used as "flash cards."

An additional feature of the ***Study Guide and Workbook*** is the discussion in the **Instructions and Suggestions** section on study and review techniques and test-taking skills and strategies using sample questions to assist students in effectively reading and interpreting questions to determine the correct answers.

Answer Keys are provided in the Appendices.

Acknowledgments

We thank the numerous adopters of ***An Illustrated Atlas of the Skeletal Muscles*** for requesting the development of an accompanying ***Study Guide and Workbook***, and also Douglas Morton and Dona Mendoza for their support in this project. As Project Manager, Dona Mendoza kept everyone on task and on target, attending to the innumerable details and coordinating work between the authors and Joanne Saliger, the compositor. We also thank Peggy Firth at Firth Studios for permission to use the skeletal and muscle illustrations from An Illustrated Atlas of the Skeletal Muscles.

We also wish to express our gratitude to the following individuals who reviewed the manuscript and provided useful comments:

Michael Bachus, Oklahoma Health Academy Massage Therapy Program
Marirose T. Ethington, Genesee Community College, Batavia, NY

Contents

Instructions and Suggestions on How to Use the Study Guide and Workbook

The *Study Guide and Workbook* includes important points, key ideas, charts, diagrams, and other information that will help you understand the skeletal, muscular, and muscle–nerve relationships more thoroughly. The diversity of objective and written questions, labeling and coloring opportunities, clinically based questions, and practical hands-on exercises will help you meet those objectives. Repetition of topics in questions and exercises in various formats will reinforce your understanding and working knowledge of that information.

Following are some suggestions, recommendations, and hints that you may find useful for effective class participation, review, and test-taking techniques and strategies, as well as posttest assessment of your performance and progress.

Lecture and Laboratory Activities

Preparing for quizzes and tests starts with regular attendance and active participation in class and the laboratory. Although recording lectures may be permitted and lecture notes may be distributed on paper or electronically, take good written notes. You will listen more intently and critically and will have a better chance of noting instructor emphases on various topics that usually provide useful hints of what will be stressed on lecture and laboratory quizzes and tests. Don't hesitate to ask questions to clarify material you don't understand, and volunteer answers to questions posed by your instructor and other students. Also take the opportunity during office hours to get to know your instructor and obtain individual help. Learning is being involved!

Studying for Quizzes and Examinations

Studying for the next quiz or test begins right after the previous one. Every test and quiz prepares you for the next one because new topics build on previous ones. Take a few minutes to go over these quizzes and tests as soon as they are returned. You want to be confident that you understand what you got correct and make sure it wasn't just a lucky guess! You especially want to identify those questions that you got wrong and determine what were the correct answers and why. Not only do you identify topics you need to clarify, it is an opportunity to assess whether your study approach requires some adjustment. Passing tests is an immediate goal; preparing for and passing tests as a regular assessment of the depth and breadth of your knowledge, and as a foundation for future courses, graduate study, and professional practice, is the long-term goal. Go for the long-term goal!

It may be a cliché, but it's a useful one: Study Early and Often! Consider this study guide as providing exams and quizzes that you can take just as if they were administered in class. Effective use of the study guide provides a diversity of the types of questions that might be asked and also provides the opportunity to evaluate the effectiveness of your study approach and knowledge of relevant materials. Don't assume these questions will be the same questions on the "real" test.

When and where you study and whether alone or with others is a matter of choice. Studying alone has the advantages of setting your own pace; determining when, where, and what you want to study; and figuring out which study methods work best for you. At the same time, there are distinct advantages to studying with another person of similar focus and purpose. A fellow student may use some strategies that you may find also work well for you. You can ask questions, provide suggestions to lead one another to correct answers, and participate jointly in the "hands-on" study exercises and Case Study questions. Either way, it is essential that you, and anyone with whom you are working, have seriously and thoroughly completed the topics to be reviewed. Study your text and regularly review class notes and laboratory activities. An initial scan of the study guide will provide an overview of the types of questions and ways to look at relevant information when you encounter it. You will very likely find yourself developing your own practice test questions. Learning is being involved!

Quiz yourself when there has been a lapse of time (several hours, a day or two) since you actively studied the course material. This will enable you to evaluate whether you know the material "on your own" as you will need to on an actual exam. Read the instructions for each type of question completely and carefully. Answer those questions about which you are sure; go back to those questions about which you were initially unsure. Do not look up answers to individual questions as you answer them. Subsequent questions often provide hints about previous questions. Also you may see answers to later questions, thus undercutting the purpose of your pretest study—determining whether you know the material "on your own." Grade yourself after answering all the questions selected for your "quiz" or "test" just as your instructor would. Then go over your study guide quiz or test and recheck what you answered correctly (again, make sure you really understand why it was the correct answer and not just a lucky guess!) and determine the correct answers for wrong responses. Later go back to the relevant course materials and reread those topics you need to clarify for yourself. If you write your answers in the study guide, make sure that you have a card or paper to cover up the answers so you can effectively re-quiz yourself on these questions at a later time.

To study effectively, you need to make meaningful connections between topics; this moves your understanding to a broader and more comprehensive level. Memorizing class notes through repeated reading of them and the use of mnemonics are two techniques that many students use. One potential advantage is that they may provide you with quick recall of some types of information. The primary disadvantage is that you may forget a critical part of memorized class notes or part of a frequently long mnemonic letter/word sequence. For example, a common mnemonic for the wrist bone sequence is Sally (scaphoid) Left (lunate) The (triquetral) Party (pisiform) To (trapezium) Take (trapezoid) Cathy (capitate) Home (hamate). An even longer one for the 12 cranial nerves is On (olfactory) Old (optic) Olympus (otic) Towering (trochlear) Tors (Trigeminal) A (abducens) Finn (facial) And (auditory) German (glossopharyngeal) Viewed (vagus) Some (spinoaccessory) Hops (hypoglossal). Making a set of flashcards may be useful. Although they may be initially useful, memorization, and use of mnemonics should serve mainly as steppingstones to study approaches that lead to long-term comprehension, retention, and practical understanding of course materials.

Test/Quiz Taking Techniques and Strategies

Before answering any questions, quickly scan through the test or quiz to note question formats, content, and point values and distribution. Carefully read the directions for each type of question. Note the questions on material with which you are familiar and confident. Which questions you select to answer first (easy, hard, objective, written, low-point questions, high-point questions) is a matter of choice. Proceed through the test or quiz and answer all the questions that you know; bypass those you don't. As you go through the test, you may find information that provides clues to previous questions. When you have finished the test, quickly review it to confirm that you have answered all questions and have done so correctly. You may find some for which you selected or wrote an incorrect response, but unless you actually did make a mistake, **don't change answers.**

Labeling Illustrations

First read the directions to understand what information is requested and where it should be placed on the page. Visualize the bones, muscles, or body movements to assist you in providing the correct information. Write the names of labeled structures and/or other requested information about the labeled structures in the corresponding spaces.

Matching

In Matching Questions, select the correct lettered term(s) from the list of choices that correspond(s) to the question, phrase, or word with which items are to be matched. Read the directions carefully before starting because Matching Questions are not necessarily one-to-one pairings; some answer choices may be used once, more than once, or not at all.

In the blank space to the left of each descriptive phrase, write the LETTER(S) of the corresponding bones.

1. _____ the "forehead" bone
2. _____ the lower jaw bone
3. _____ the bone containing the ear bones
4. _____ the "cheek" bone
5. _____ TWO bones containing air sinuses

A. Zygomatic
B. Occipital
C. Maxilla
D. Parietal
E. Frontal
F. Temporal
G. Mandible

Question Analysis Note that each question requires one answer, except for #5, which requires the names of two bones. You might also notice that the questions are about bones on the front and side of the skull. Some lettered choices won't be used, and some may be used more than once. You should first read each question as a fill-in-the-blank question to determine whether that is one of the choices. For example, **The "forehead bone" is the ______.**

The correct answers are: **1.** E **2.** G **3.** F **4.** A **5.** C, E

Statement Completion

Select and **circle** the term in parentheses ( ) that correctly completes the statement. Note that in some questions, you are required to circle a choice from each of two sets of terms in parentheses. Rephrase the question with the possible answer(s) to assess which one(s) sound(s) correct.

1. All anatomical terms applied to the human assume the person is in anatomical position—that is, standing up, facing forward, with the palms of the hands directed (**anteriorly, posteriorly**).

Question Analysis The question asks in what direction the palms of the hands are held in the anatomical position. Rephrase the question: "The palms of the hands are directed __?__ in anatomical position." Then ask yourself which of the two choices is correct.

The correct answer is anteriorly.

2. Two muscles with actions that are (**opposite, the same**) would be referred to as (**antagonistic, synergistic**) muscles.

Question Analysis When you are asked to select a choice from each of two sets of terms, you also rephrase the question as a fill-in-the-blank format and complete with each pair of alternatives to determine what sounds correct. **In this case, there are two correct responses: opposite and antagonistic and the same and synergistic.**

Multiple Choice

In Multiple Choice questions, you select the correct or best answer from a list of lettered choices. Multiple choice questions can also be considered a list of True–False choices from which you select the one choice that completes a valid statement with the question stem. **First**, read the question completely and carefully, noting the key terms to determine what the question is asking. **Second**, cover or ignore the answer choices and consider what would be a correct answer if the question were a fill-in-the-blank format, and check whether your answer is one of the lettered choices. **Third**, if your answer wasn't one of the choices, read the question stem with each choice and eliminate obviously wrong or implausible choices and select the one that sounds correct.

1. _____ Which one of the following terms is used for both an articulation surface as well as a surface for muscle attachment?
   A. Facet
   B. Fontanel
   C. Foramen
   D. Fossa

Question Analysis You may know it right away from asking "What is a term used for both an articulation and muscle surface?" But if not, you may recognize it from the choices or come to an answer by a process of elimination by recognizing to what each of the other choices refer. **The correct answer is D. Fossa.**

Fill in the Blank

Write the appropriate word or phrase in the blank space or spaces to complete a valid statement. First read the question completely to note whether one or more than one written response is required (often separated by commas, which can be missed). Look for key terms. Read or rephrase the question with the possible responses to assess whether your answer sounds correct.

1. Listed from superior to inferior, the three sections of the sternum are the ____________________, ____________________, and ____________________.

Question Analysis Most important, note that the question asks for the three sections of the sternum to be listed from superior to inferior, not just what are the three sections of the sternum. Answer in the specific sequence requested, or your response will be incorrect.

The correct answer is: manubrium, body (or gladiolus), and xiphoid process.

Short Answer

First read the question completely to note exactly what information is requested. Note if more than one type of information is required. Write your answer in brief sentences unless a list is requested. Rephrase the question with your answer to assess whether your response sounds correct.

1. Describe two basic differences between synarthrotic and diarthrotic articulations.

____

____

Question Analysis You are asked to describe **two differences between synarthrotic and diarthrotic articulations.** *Do not just list an example of each type of articulation.* That is the most common mistake students make when asked to describe differences between two things. The question does not specify the particular types of differences, so this gives you choices. First, you have to remember to what each type of articulation refers and what are their specific structural and functional characteristics. Regardless of how you write your answer, remember to describe the alternate conditions *in both structures*.

Thus, note that in the following sample answer, differences in both structure and degree of movement permitted are clearly indicated.

In synarthrotic articulations, the bones are closely connected to one another by fibrous connective tissue that allows little movement between the involved bones. In contrast, in diarthrotic articulations, the involved bones are separated by a fluid-filled capsule that permits considerably more movement between the bones.

Case Studies

You will read a short paragraph that describes the details of an injury, disease, or condition on which one or more follow-up questions are based. First read the paragraph completely and note the important information about the individual's condition. You might want to write down relevant concepts and terms. Answer each question in full sentences if required. Reread your answers for accuracy, completeness, and clarity.

A 13-year-old boy and his father, the high school basketball coach, go to a doctor, both complaining of a "sore knee." The father explains that they constantly practice the quick changes in direction and jumping movements typical of basketball. During the examinations, the doctor notes that the location of the pain and soreness are between the knee cap and leg on both, but at slightly different positions. It was just below the knee cap in the father, but close to the lower leg bone in the son.

1. What are the technical names of the knee cap and lower leg bone?

2. What are one or two injuries between the knee cap and lower leg that could account for the similar "soreness" in both the father and son?

3. Considering the particular location of the soreness, what major muscle group would be involved in both the father and son?

4. Considering the slight but specific difference in location of the pain together with the ages of the father and son, explain what specific problem had developed in each.

Question Analysis The similarity in symptoms could suggest a common condition, but the difference in age of the patients and in the subtle but specific location of the soreness in each are important issues to consider.

Suggested Answers

1. The patella and tibia, respectively
2. Inflammation or tearing of the tendon (sometimes referred to as a ligament) between the lower edge of the patella and the tibia
3. The quadriceps muscle group
4. The son would have areas of incomplete ossification "surface features" of bones where muscle tendons or ligaments attach, compared with the father. The excessive force of the quadriceps muscle contraction in the father could have caused subpatella inflammation (tendinitis) and possibly tendon tear (tendinopathy). The closer location of soreness in the son is probably due to partial separation/loosening of the tibial tuberosity because of persistent nonossified cartilage between the tuberosity and the tibia. This condition is referred to as the Osgood-Schlatter syndrome.

When you have finished the test, review it quickly to make sure that you have answered all questions.

The Skeleton and Fractures

1

At the end of this chapter, you should be able to

1. Identify the bones of the body.
2. Differentiate between the axial and appendicular divisions of the skeleton and identify the components of each.
3. Identify the "bony landmarks" on individual bones.
4. Name selected skeletal abnormalities.
5. Identify the major types of fractures.
6. Identify types of youth fractures at the epiphyseal plates and apophyses.
7. Name three differences between the male and female pelvis.
8. Palpate various bones and their "bony landmarks" on yourself or a partner.

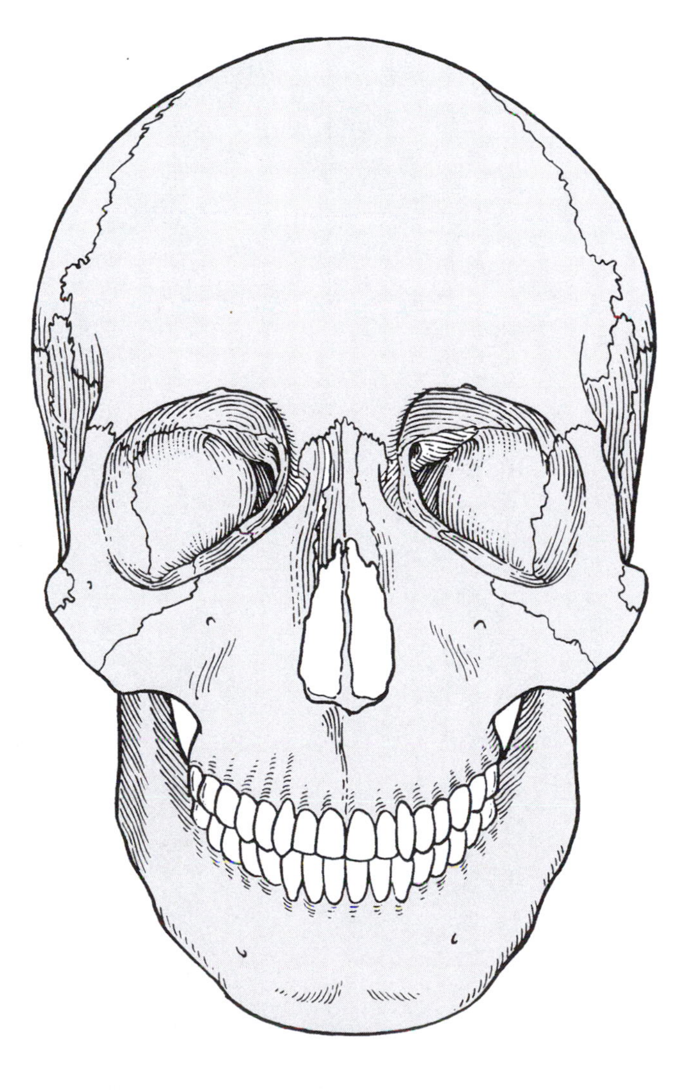

Word Search Puzzle

INSTRUCTIONS: Find and CIRCLE each of the listed terms in the Word Search Puzzle.
(Terms may read from left to right, right to left, up, down, or diagonally.)

M B C X S A G I T T A L X C C E R V I C A L
U P F D U K F P R Y Z P F U I L X C L B P K
N O N A R C E L O C X X Z B K T I B I A X S
R T V V X Y Z A C R O M I O N A A W U Z Z T
E R X A P Z K B H E X Z Y I Z L K M M Q Q R
T A Z X L P C K A S G F K D M U Z A O Q Q E
S N Z X K E D B N T D P U B I S Q L Z G Q S
Q S Q X F C C E T A M A H S Z Q U L N A Y S
A V U L S I O N E Y A S T U Q R P E G Y Z Z
Q E Z S U I D A R S S A C R U M X O V X W V
K Z R A E L H C O R T Z X E X W Z L M N P Z
K S L A R I P S F P O Q Z M V W M U I H S I
Z E Y B C A X Z K X I Z Q U M N T S V V W Z
X Y C C O C P M N X D I O H P I X A S S O F

ACROMION
AVULSION
CERVICAL
COCCYX
CREST
CUBOID
FOSSA
HAMATE
HUMERUS
ILIAC

ILIUM
ISCHIUM
MALLEOLUS
MASTOID
OLECRANON
OVALE
PUBIS
RADIUS
SACRUM
SAGITTAL

SPIRAL
STERNUM
STRESS
TALUS
TIBIA
TRANSVERSE
TROCHANTER
TROCHLEAR
ULNA
XIPHOID
ZYGOMATIC

Anterior and Posterior Views of Skeleton

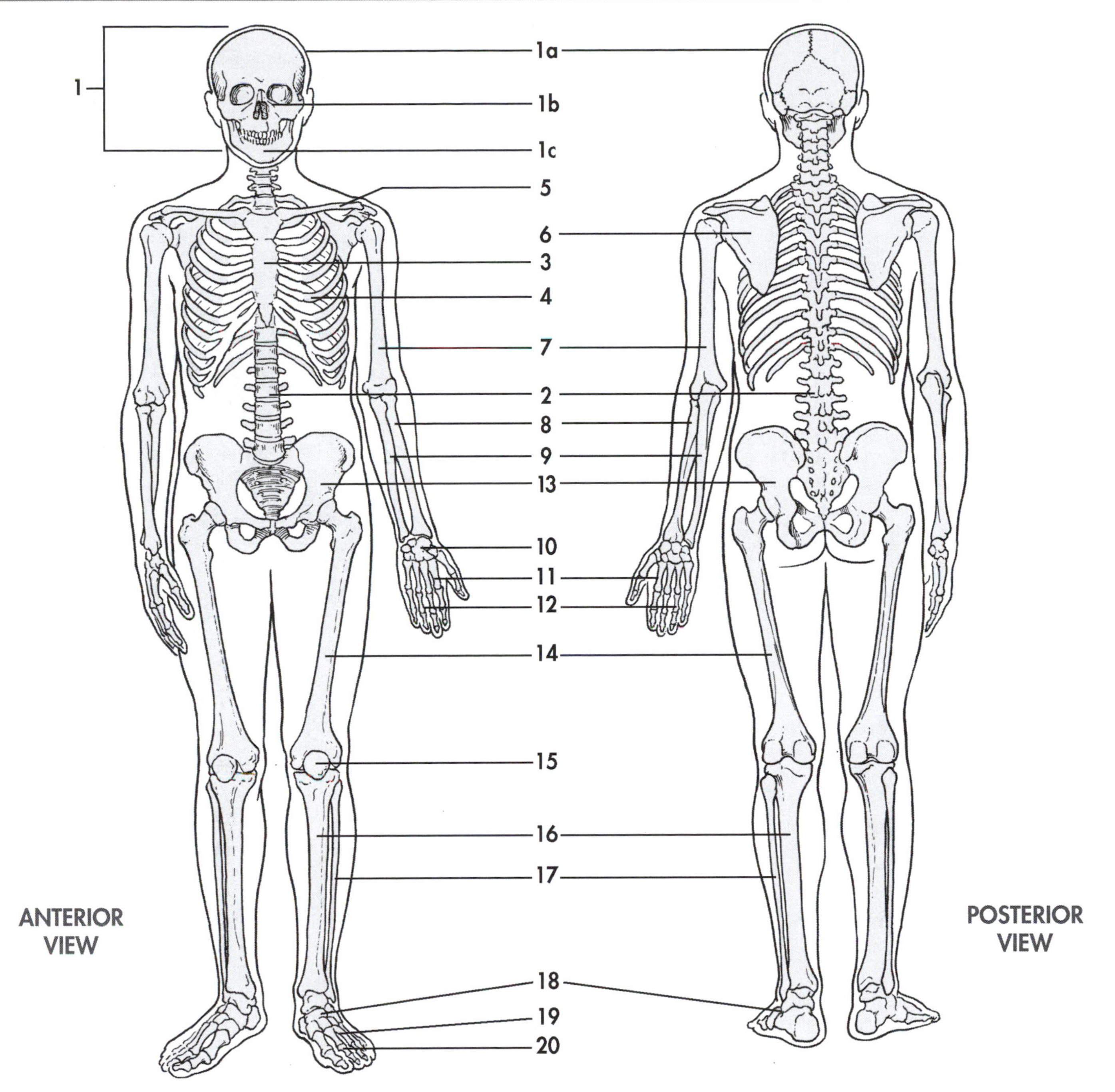

I. ILLUSTRATION IDENTIFICATION

Write the names of the numbered bones on the skeletal illustrations in the corresponding numbered spaces.

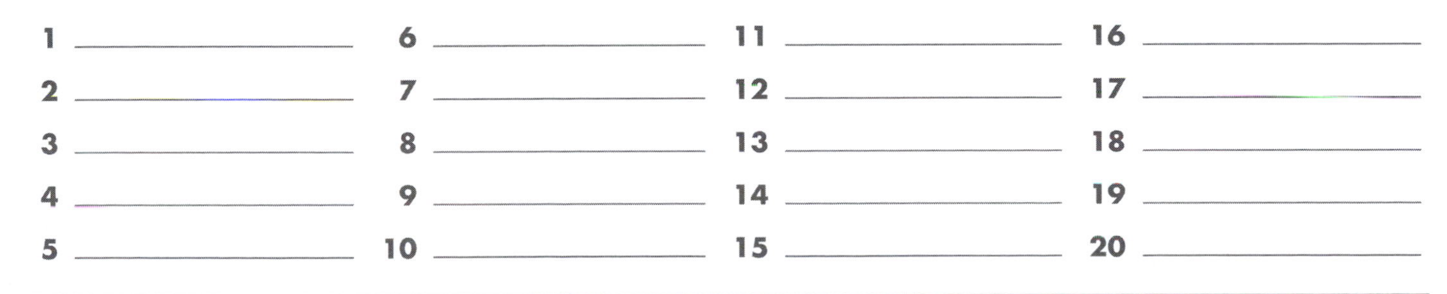

1 ________	6 ________	11 ________	16 ________
2 ________	7 ________	12 ________	17 ________
3 ________	8 ________	13 ________	18 ________
4 ________	9 ________	14 ________	19 ________
5 ________	10 ________	15 ________	20 ________

Lateral and Anterior Views of the Skull

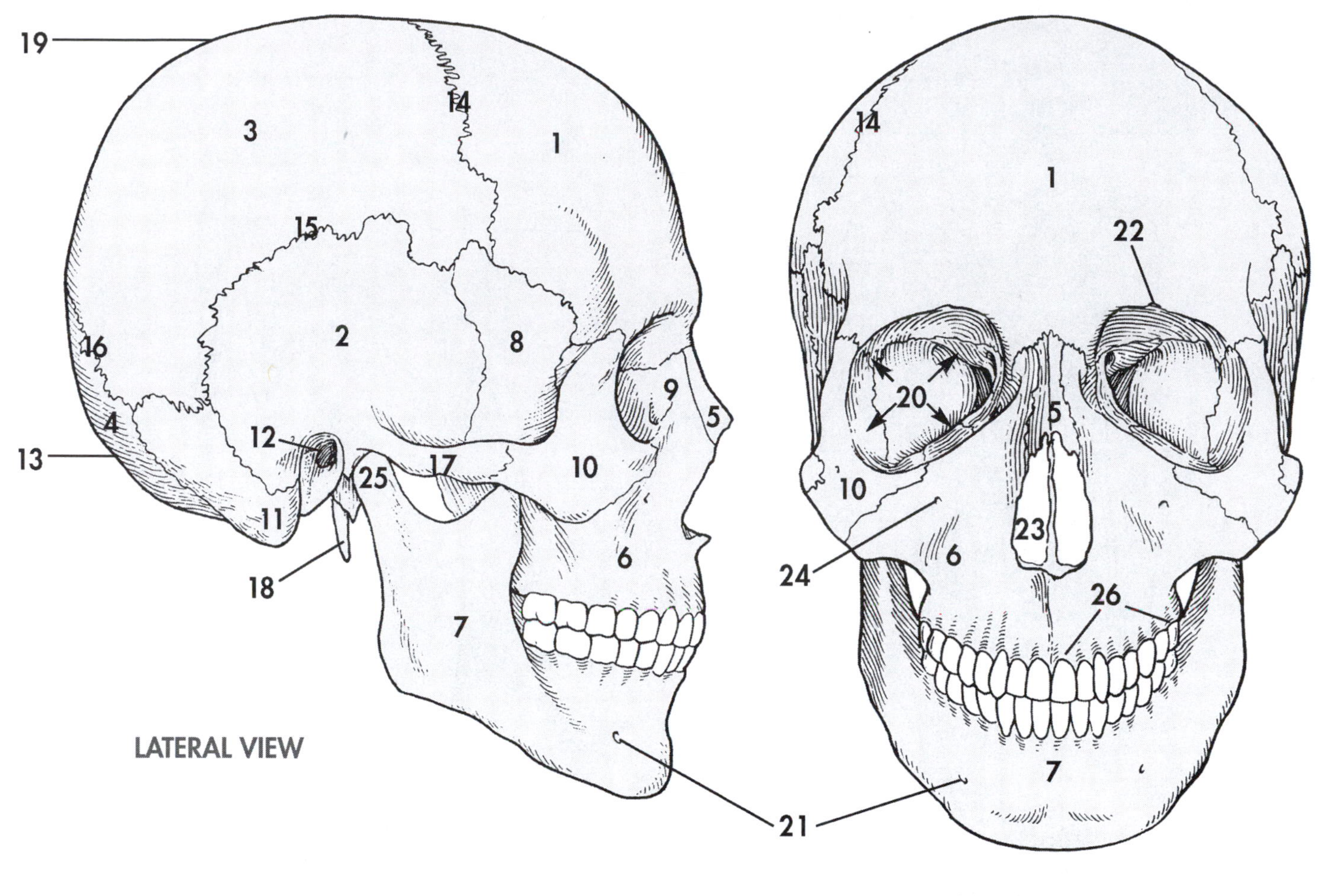

Write the names of the numbered bones and "bony landmarks" on all the following skeletal illustrations in the corresponding numbered spaces.

1 ______________________
2 ______________________
3 ______________________
4 ______________________
5 ______________________
6 ______________________
7 ______________________
8 ______________________
9 ______________________
10 ______________________
11 ______________________
12 ______________________
13 ______________________
14 ______________________
15 ______________________
16 ______________________
17 ______________________
18 ______________________
19 ______________________
20 ______________________
21 ______________________
22 ______________________
23 ______________________
24 ______________________
25 ______________________
26 ______________________

Superior and Inferior Views of Skull

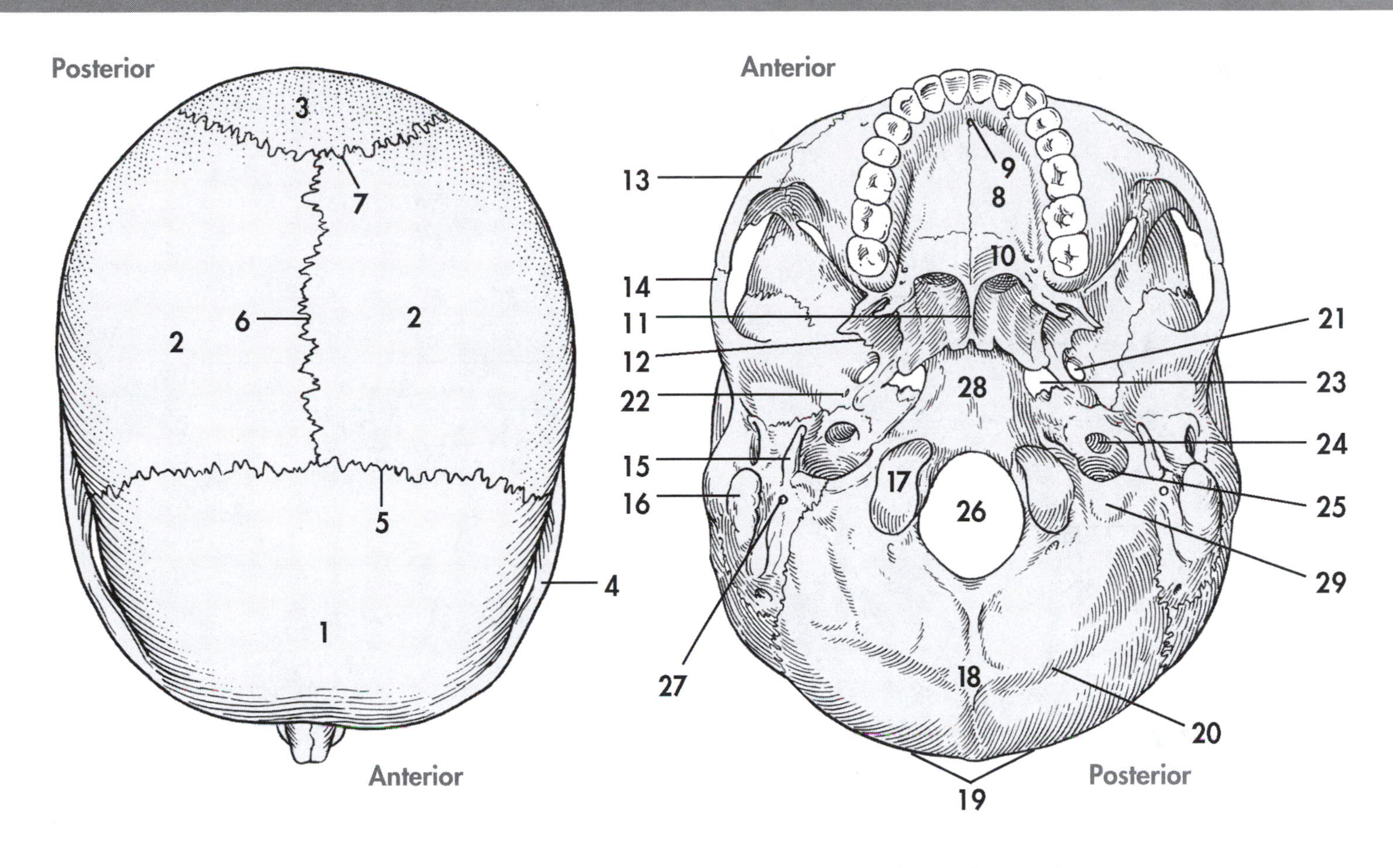

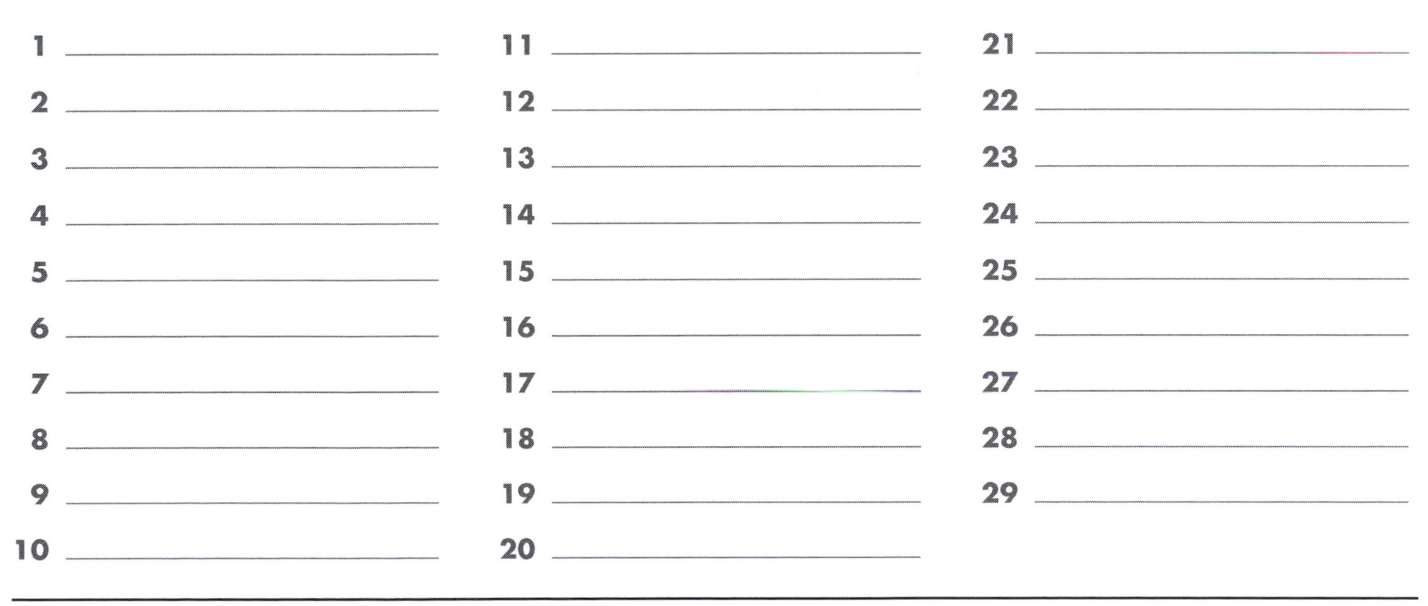

1 ____________________
2 ____________________
3 ____________________
4 ____________________
5 ____________________
6 ____________________
7 ____________________
8 ____________________
9 ____________________
10 ____________________
11 ____________________
12 ____________________
13 ____________________
14 ____________________
15 ____________________
16 ____________________
17 ____________________
18 ____________________
19 ____________________
20 ____________________
21 ____________________
22 ____________________
23 ____________________
24 ____________________
25 ____________________
26 ____________________
27 ____________________
28 ____________________
29 ____________________

Internal View of Base of Skull

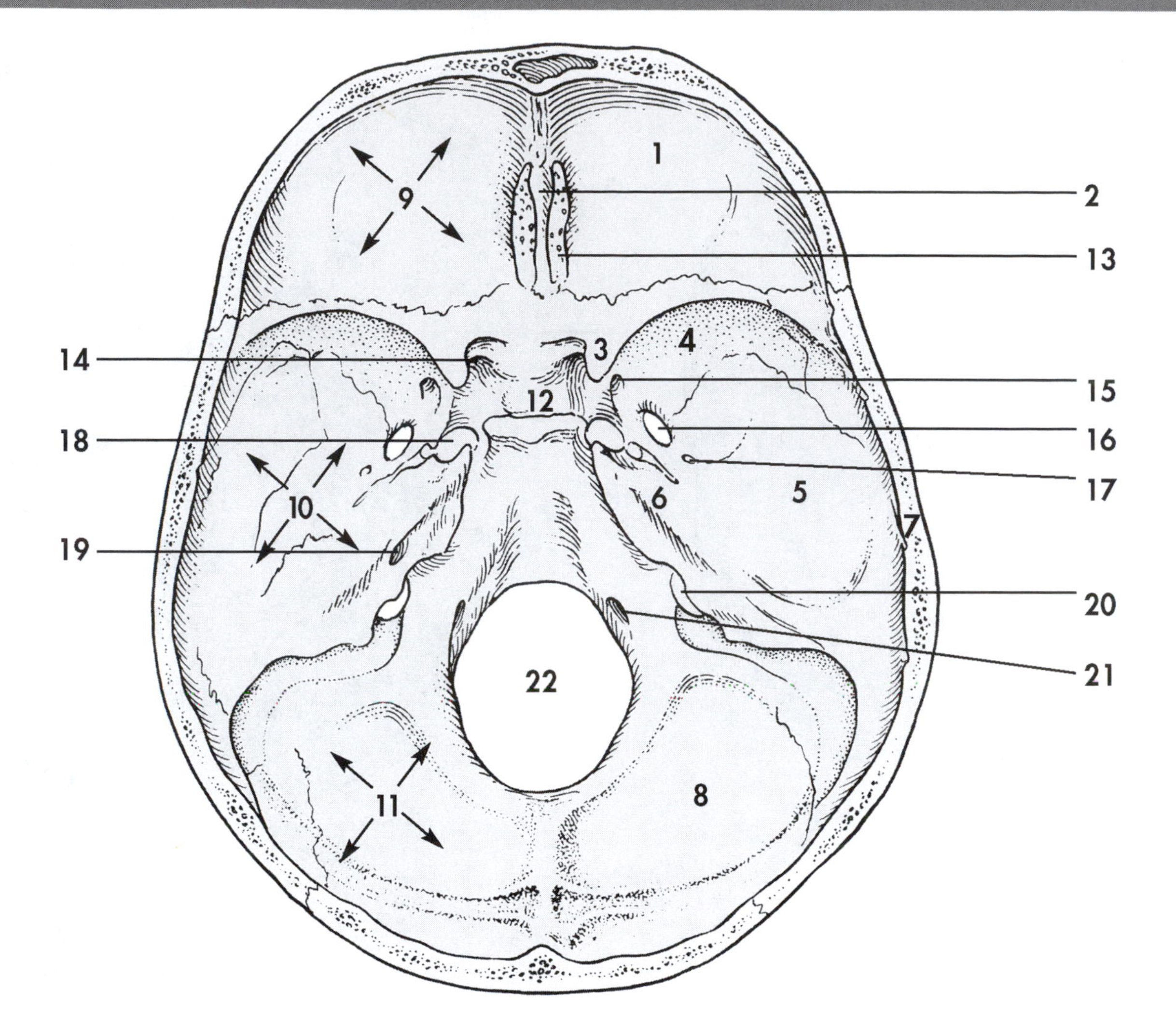

SUPERIOR INTERNAL VIEW

Mandible/Dentary

(**man**•da•ble/**den**•ter•ry)

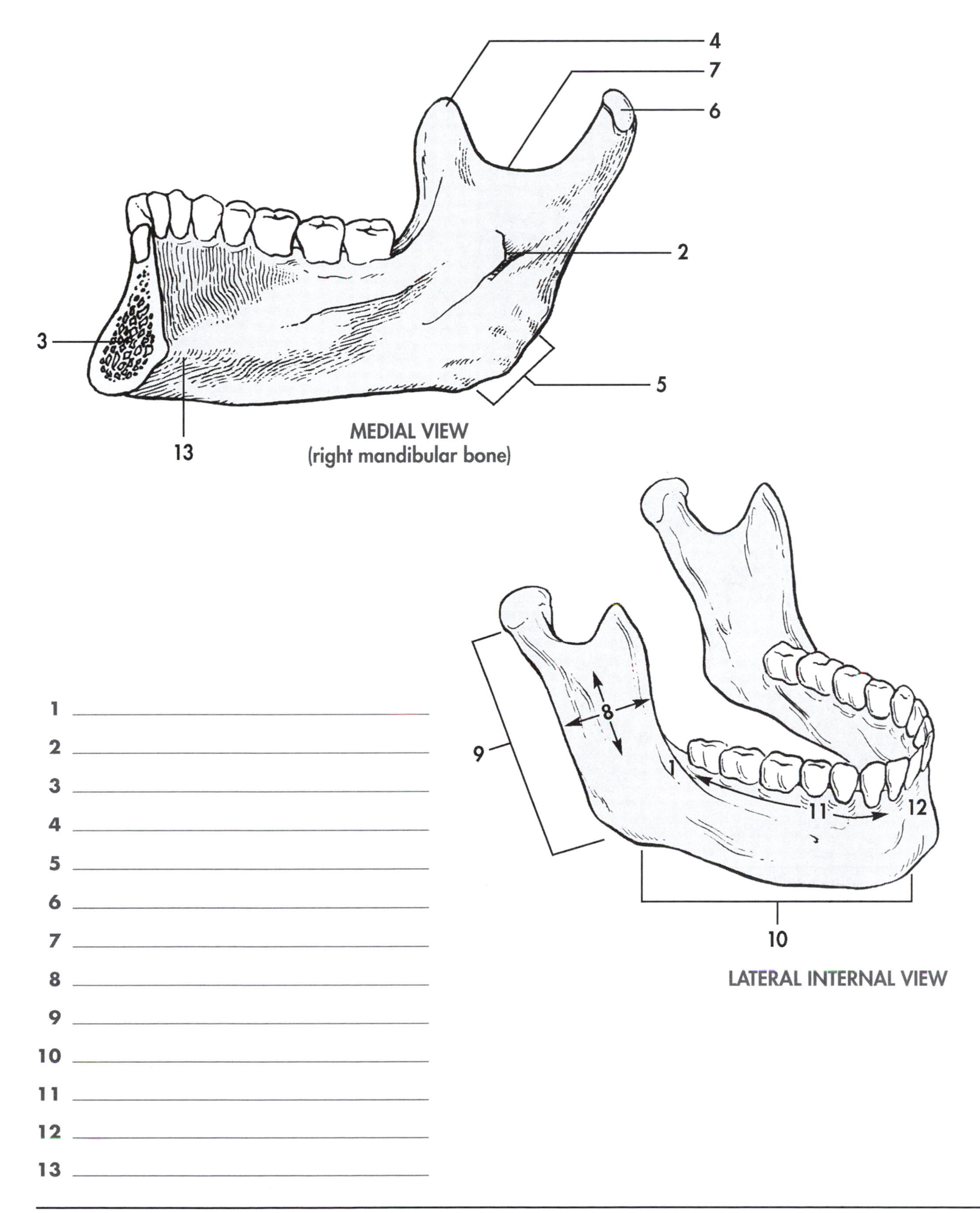

MEDIAL VIEW
(right mandibular bone)

LATERAL INTERNAL VIEW

1 ______________________

2 ______________________

3 ______________________

4 ______________________

5 ______________________

6 ______________________

7 ______________________

8 ______________________

9 ______________________

10 ______________________

11 ______________________

12 ______________________

13 ______________________

Cervical Vertebrae, Hyoid Bone, and Thyroid Cartilage

(**ser**•vi•cal) (**ver**•ta•bray) (**hi**•oid) (**thy**•roid) (**kar**•ti•lege)

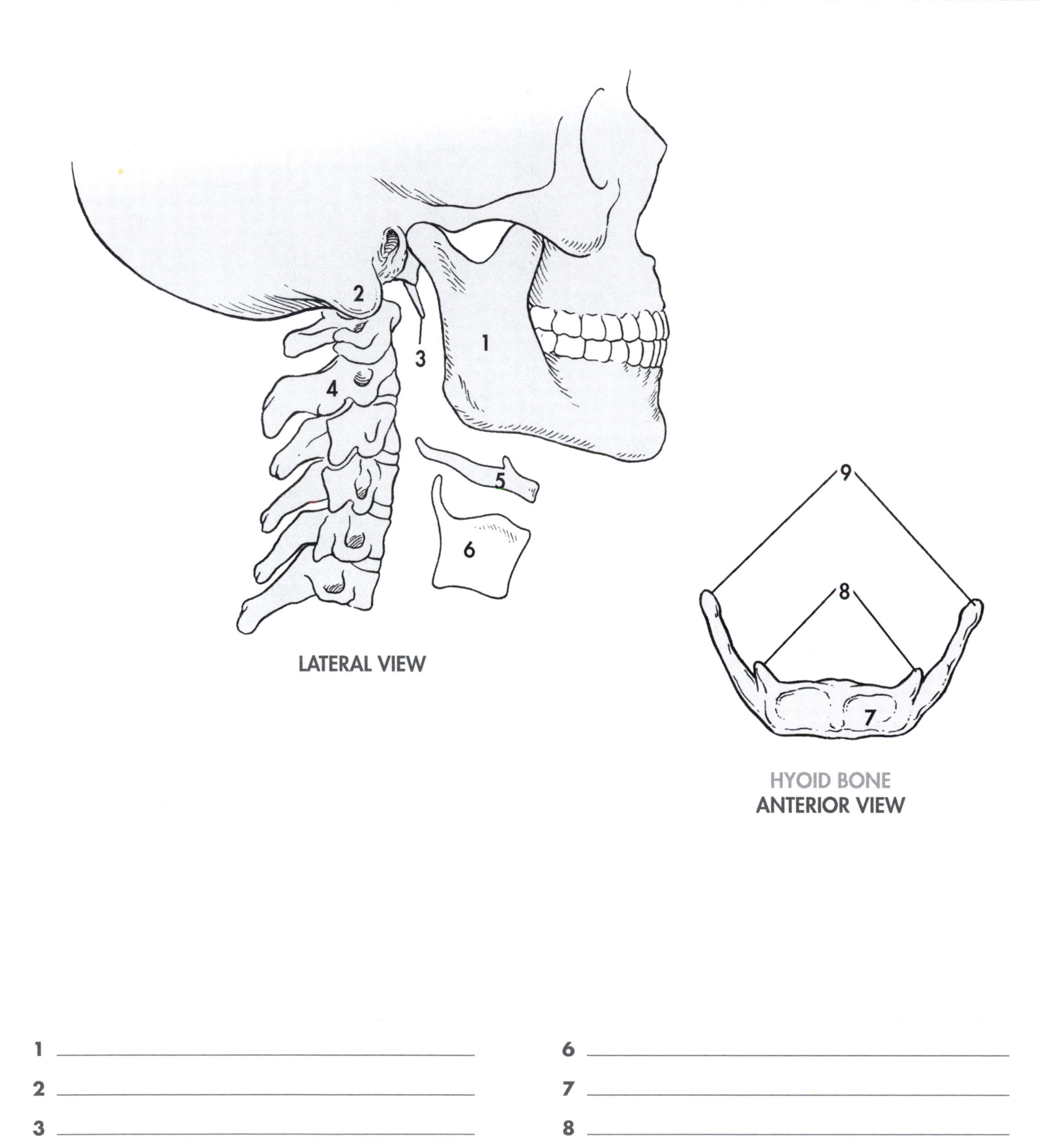

1 ______________________

2 ______________________

3 ______________________

4 ______________________

5 ______________________

6 ______________________

7 ______________________

8 ______________________

9 ______________________

Sternum and Thoracic Cage

(**ster**•num) (tho•**ras**•ik)

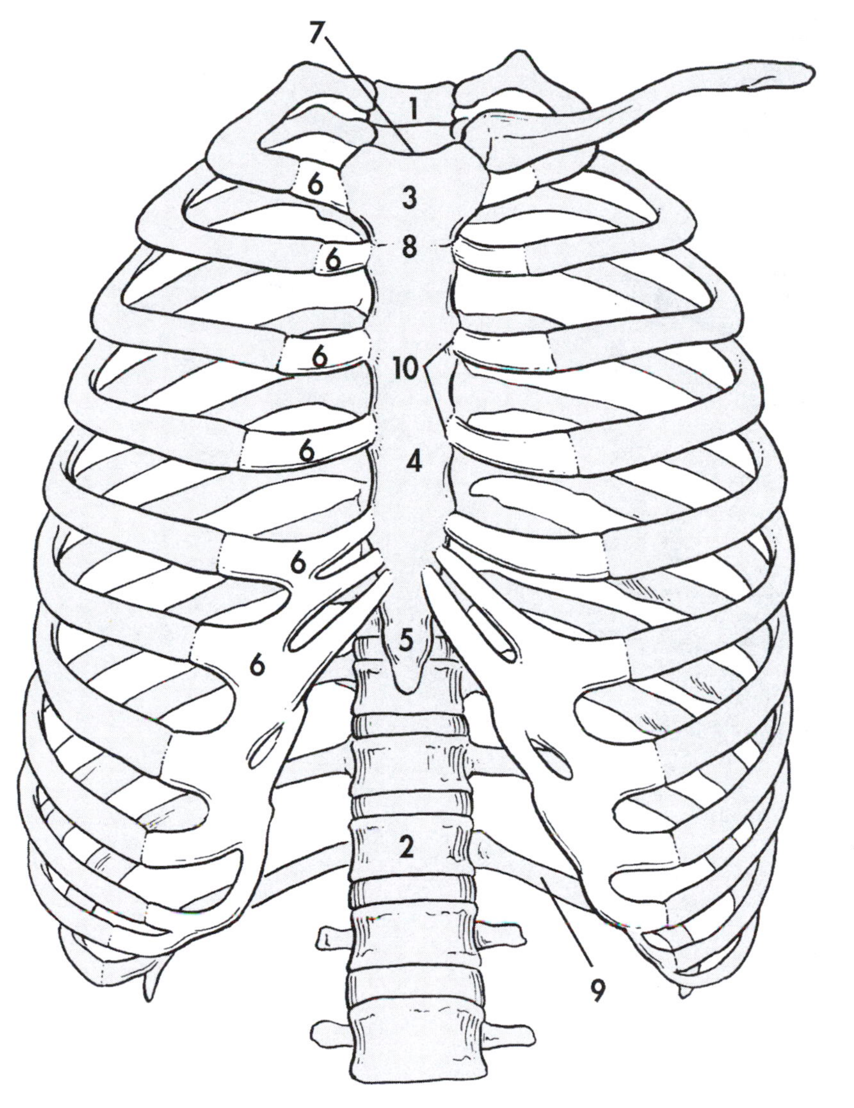

ANTERIOR VIEW

1 ______________________

2 ______________________

3 ______________________

4 ______________________

5 ______________________

6 ______________________

7 ______________________

8 ______________________

9 ______________________

10 ______________________

Ribs

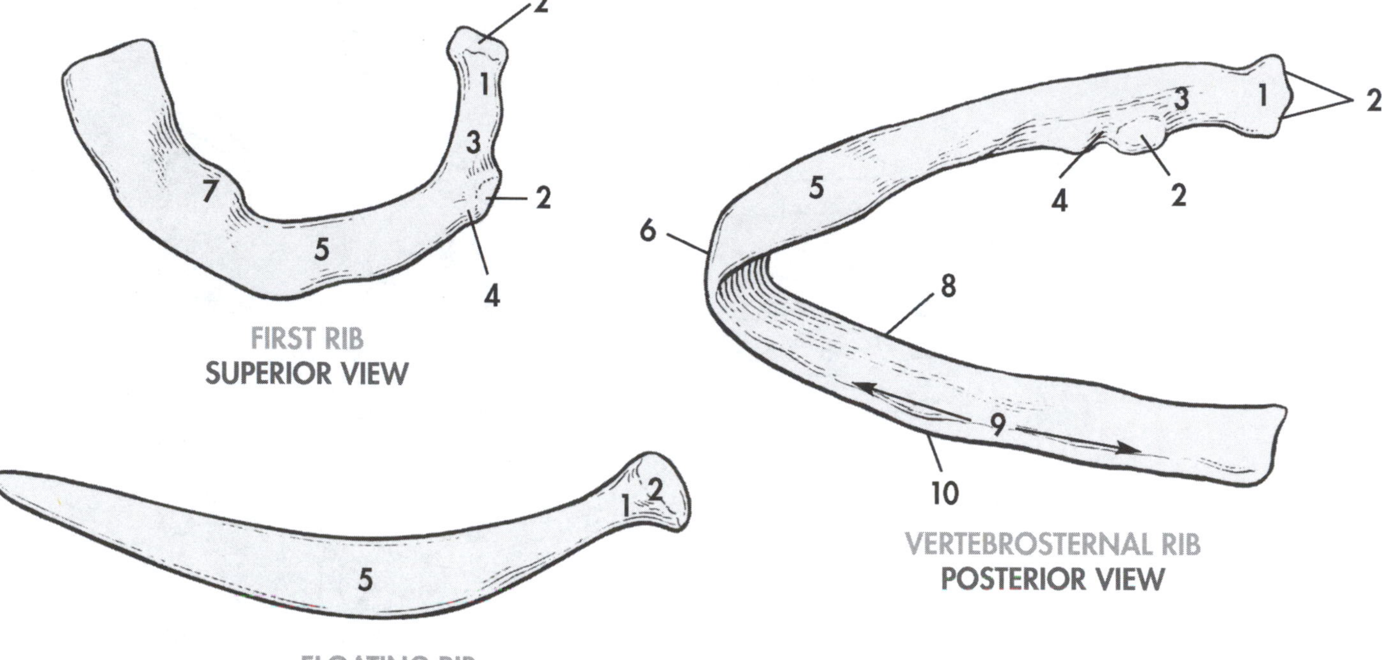

FIRST RIB
SUPERIOR VIEW

VERTEBROSTERNAL RIB
POSTERIOR VIEW

FLOATING RIB

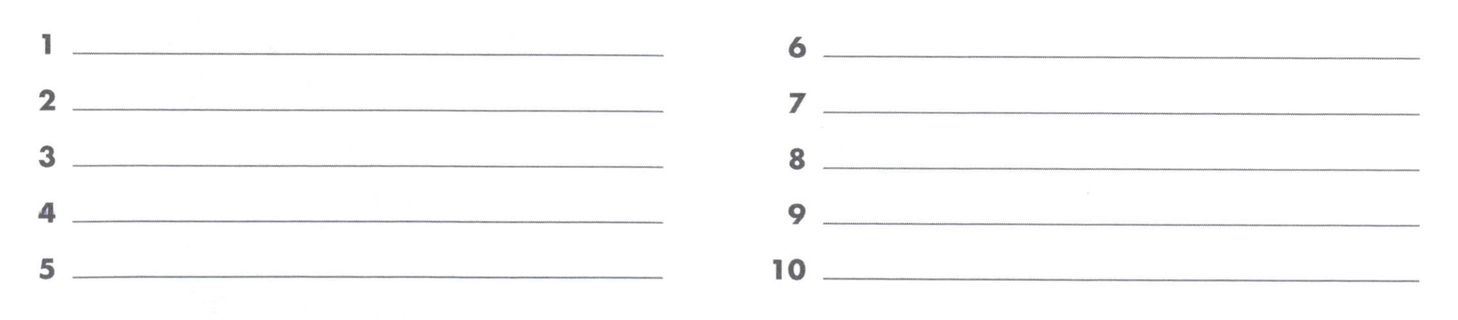

Vertebral Column

(ver•**tee**•bral)

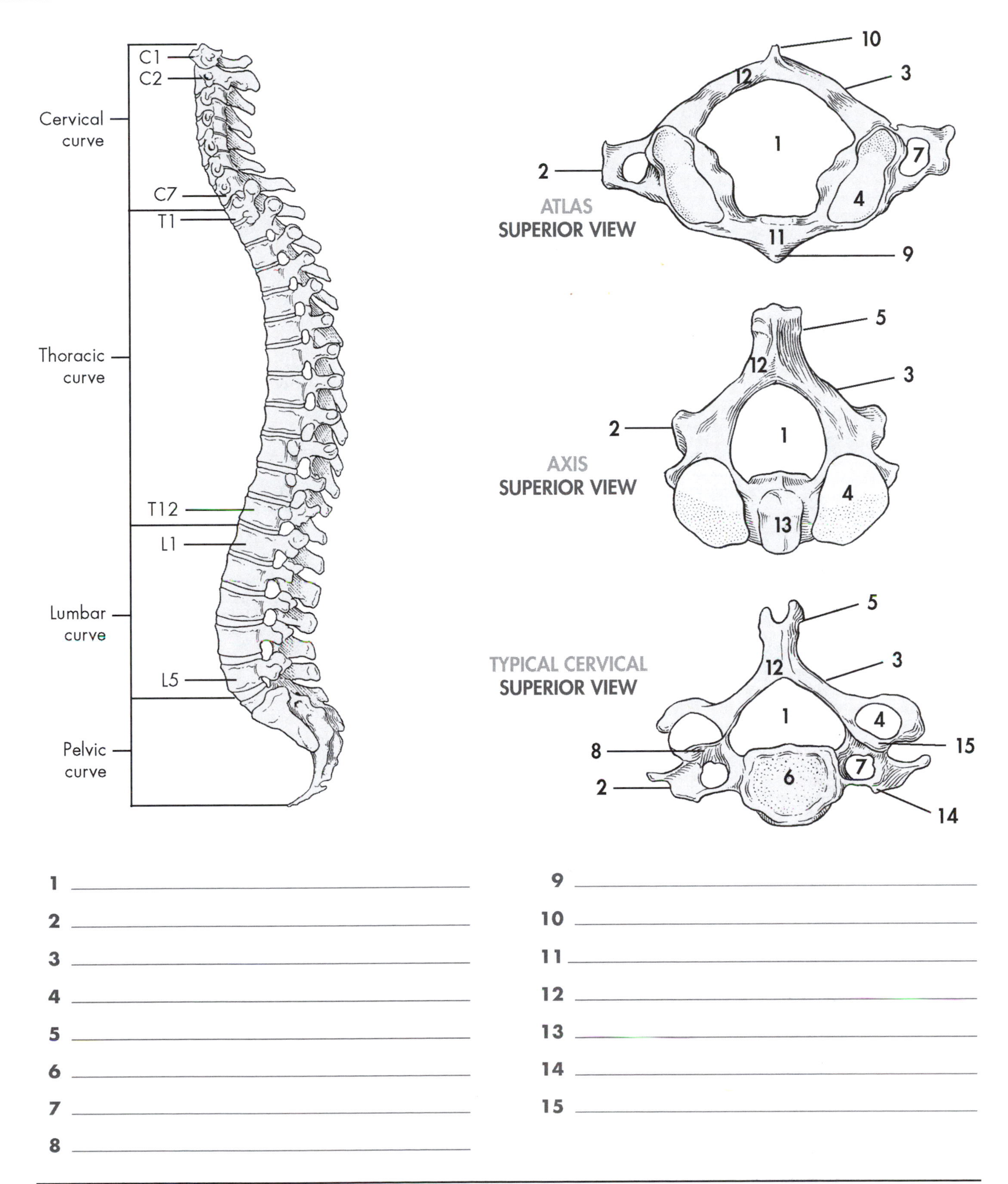

1 ______________________

2 ______________________

3 ______________________

4 ______________________

5 ______________________

6 ______________________

7 ______________________

8 ______________________

9 ______________________

10 ______________________

11 ______________________

12 ______________________

13 ______________________

14 ______________________

15 ______________________

Vertebral Column

(ver•**tee**•bral)

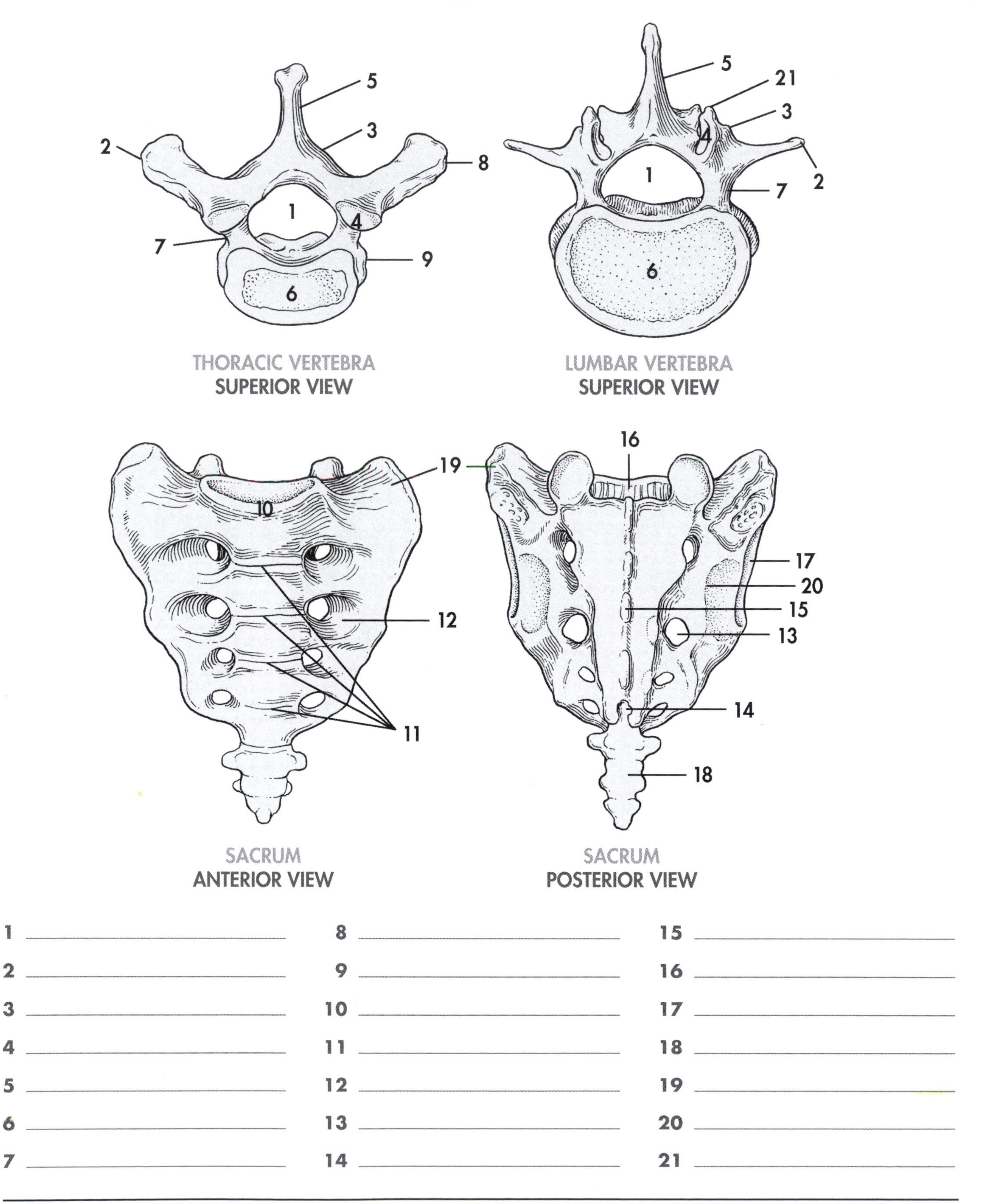

1 ____________________ 8 ____________________ 15 ____________________

2 ____________________ 9 ____________________ 16 ____________________

3 ____________________ 10 ____________________ 17 ____________________

4 ____________________ 11 ____________________ 18 ____________________

5 ____________________ 12 ____________________ 19 ____________________

6 ____________________ 13 ____________________ 20 ____________________

7 ____________________ 14 ____________________ 21 ____________________

Right Clavicle

(klav•i•kal)

SUPERIOR VIEW

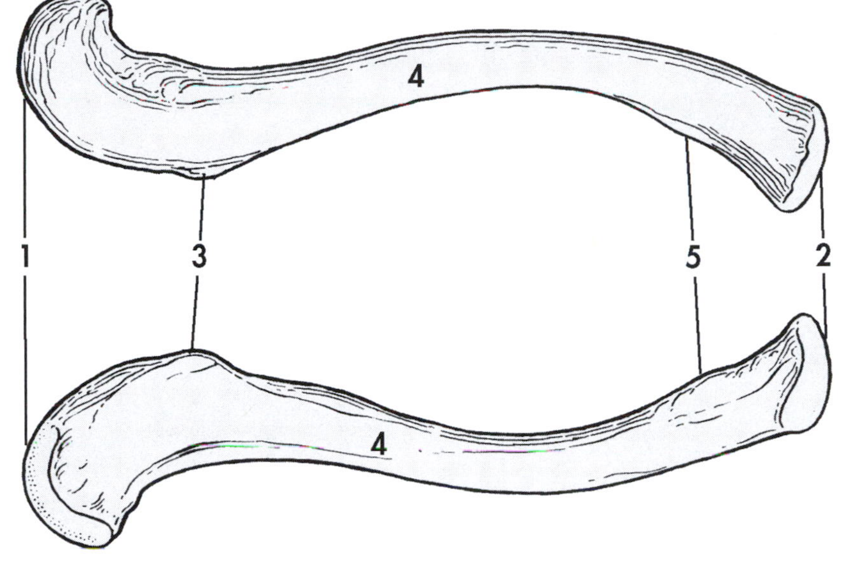

INFERIOR VIEW

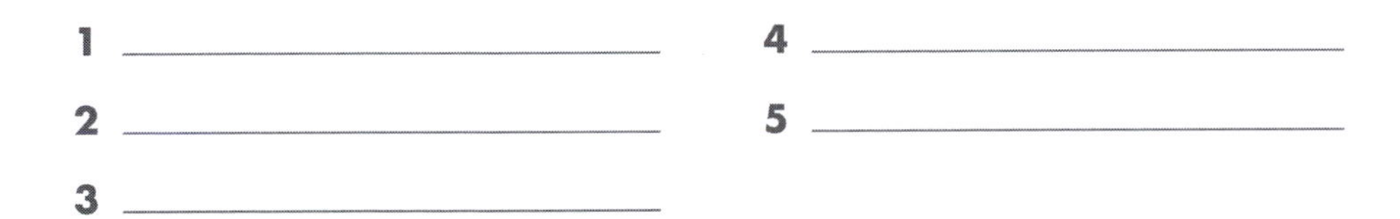

Right Scapula
(**skap**•yoo•lah)

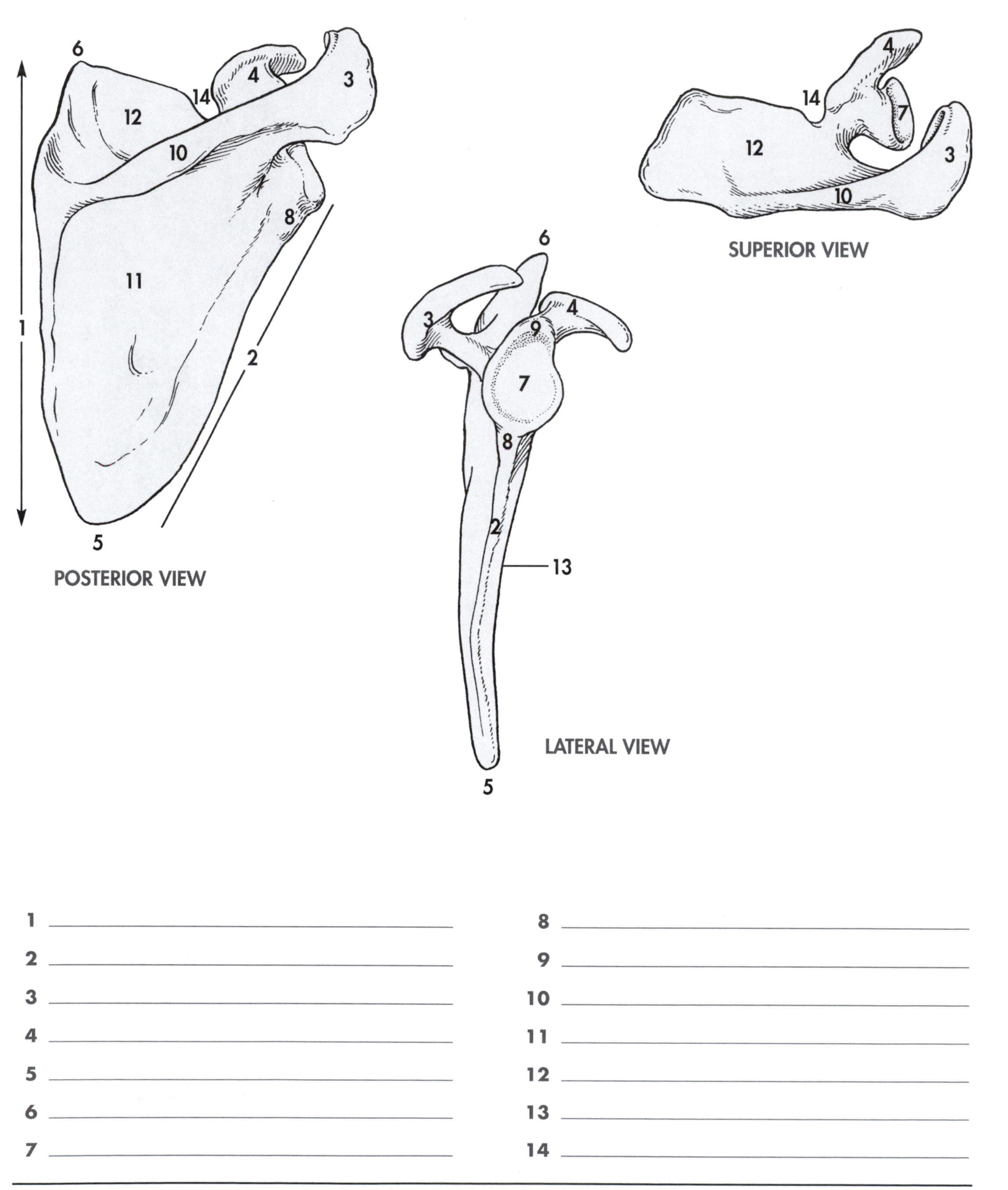

1	__________	8	__________
2	__________	9	__________
3	__________	10	__________
4	__________	11	__________
5	__________	12	__________
6	__________	13	__________
7	__________	14	__________

Right Humerus

(hyoo•**mir**•**us**)

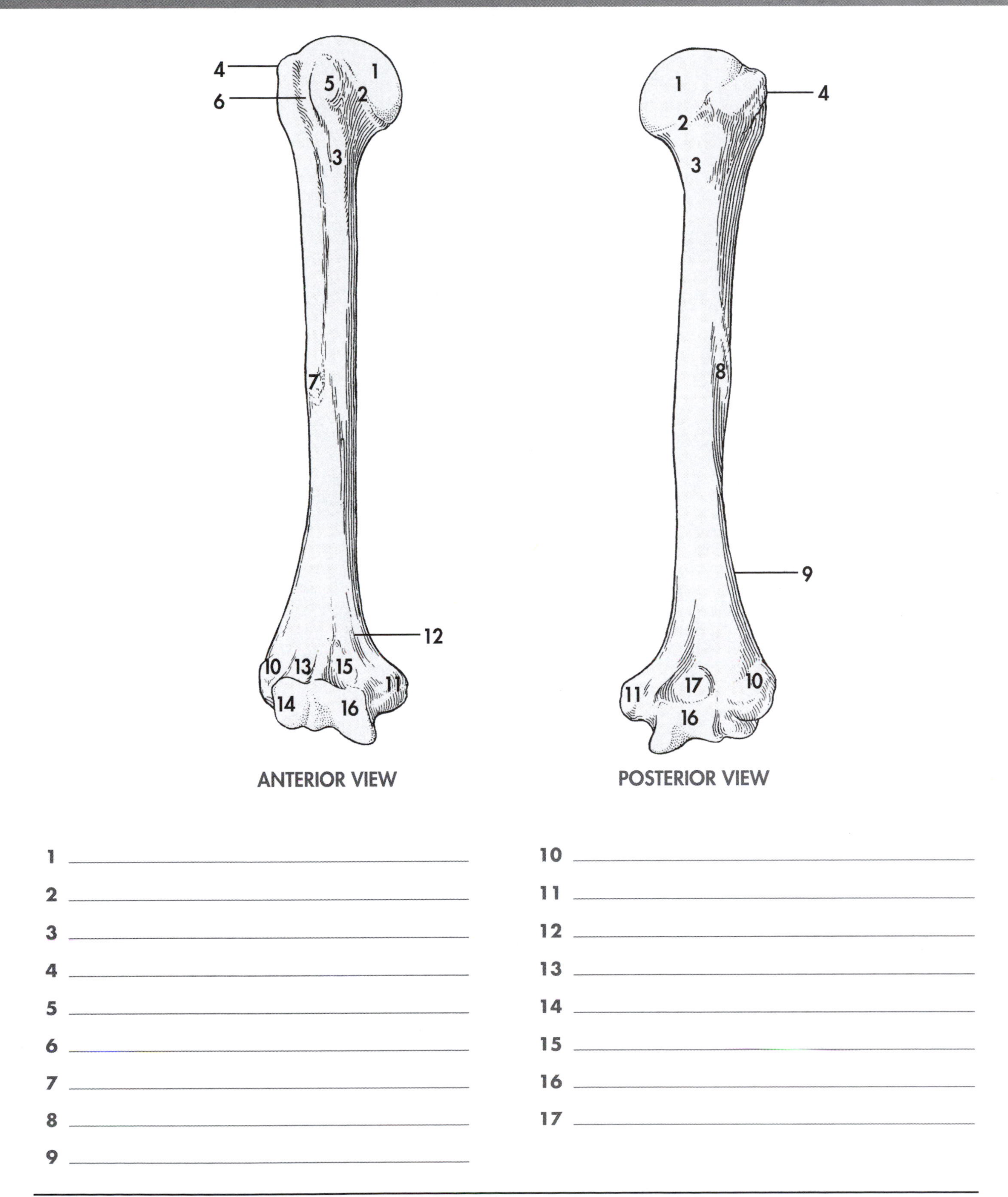

ANTERIOR VIEW

POSTERIOR VIEW

1 ______________________
2 ______________________
3 ______________________
4 ______________________
5 ______________________
6 ______________________
7 ______________________
8 ______________________
9 ______________________
10 ______________________
11 ______________________
12 ______________________
13 ______________________
14 ______________________
15 ______________________
16 ______________________
17 ______________________

Right Radius and Ulna

(**ray**•de•us) (**ul**•nuh)

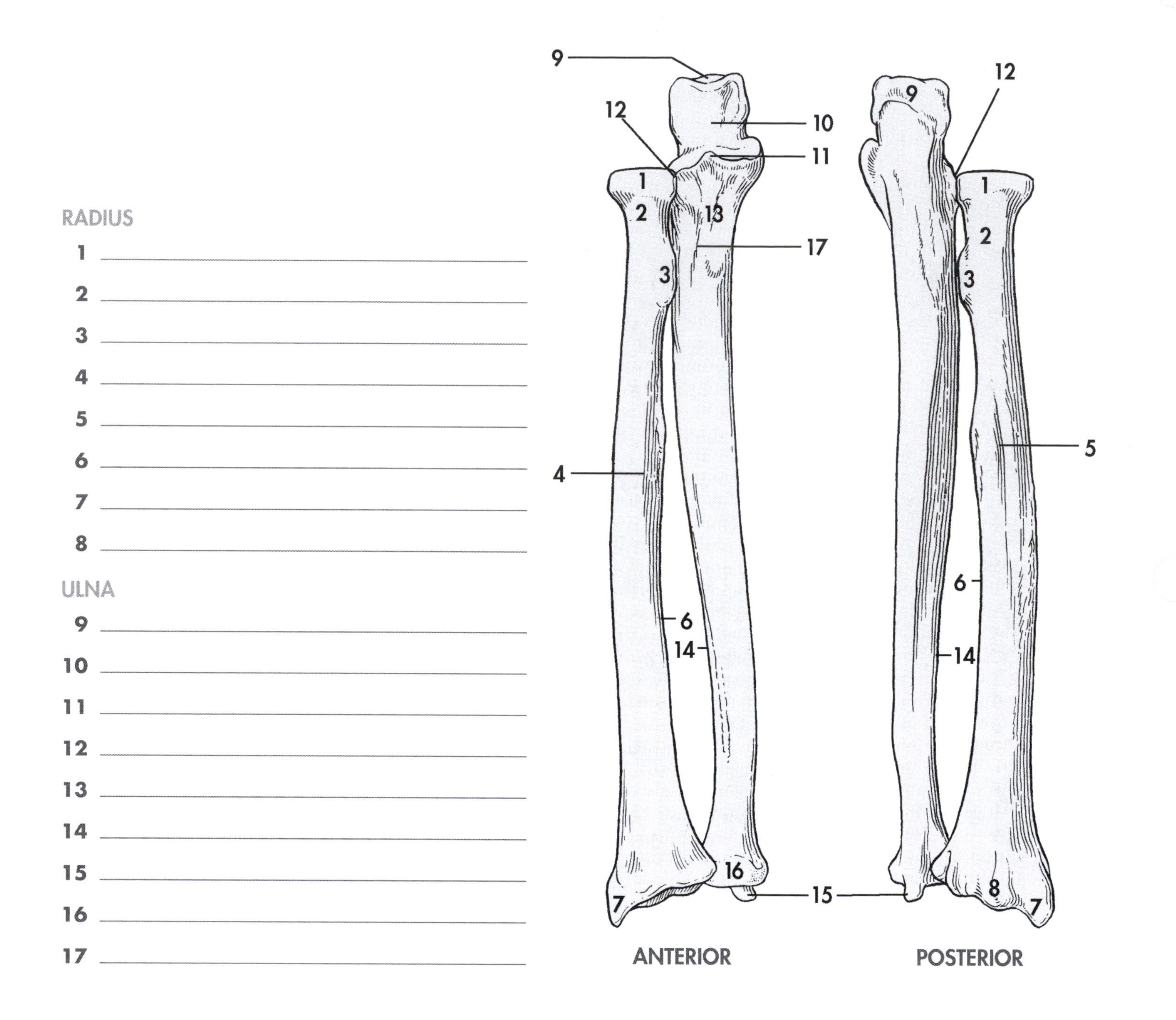

Right Hand

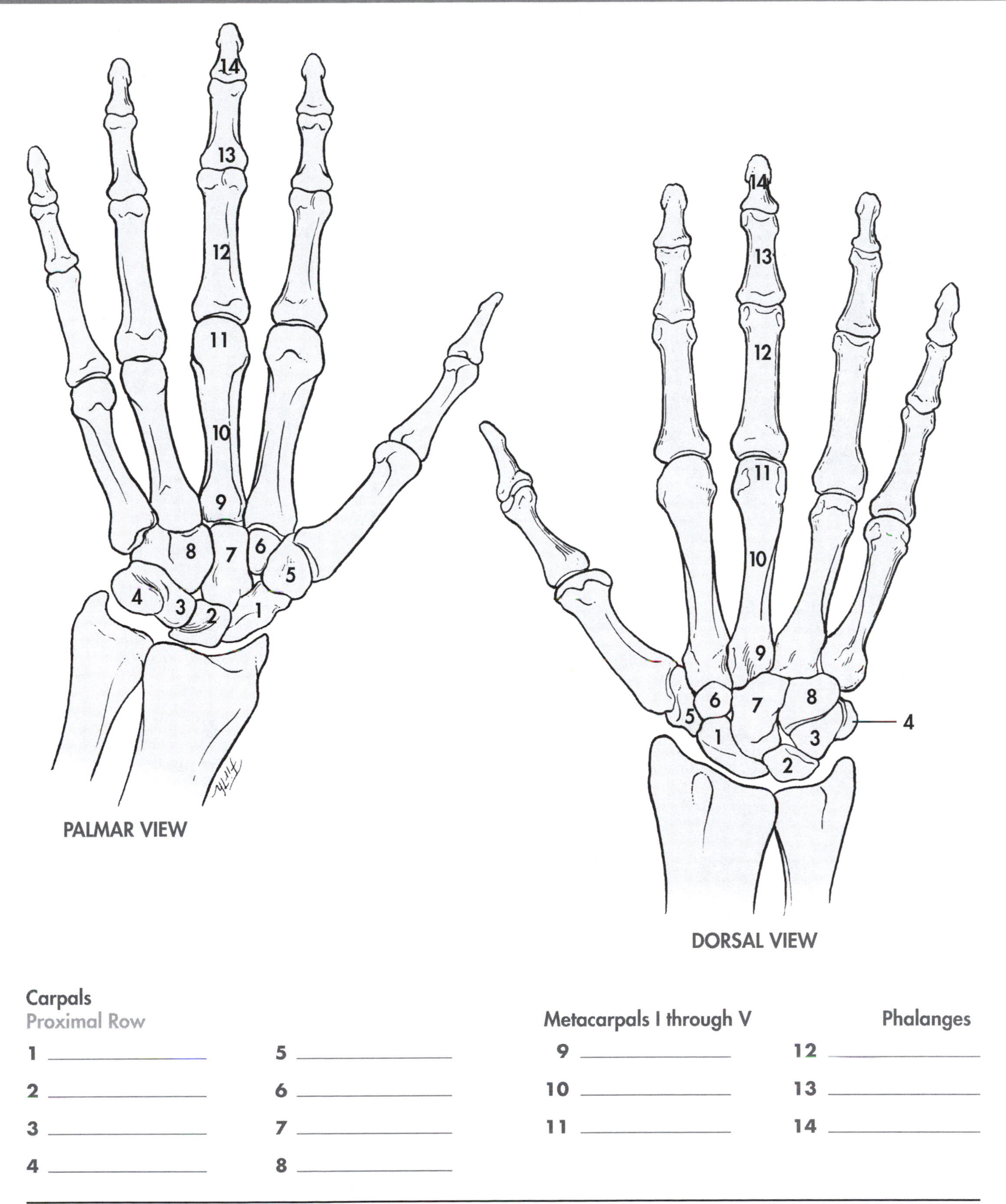

Carpals

Proximal Row

1 ____________

2 ____________

3 ____________

4 ____________

5 ____________

6 ____________

7 ____________

8 ____________

Metacarpals I through V

9 ____________

10 ____________

11 ____________

Phalanges

12 ____________

13 ____________

14 ____________

Right Os Coxa

(os **koks**•a)

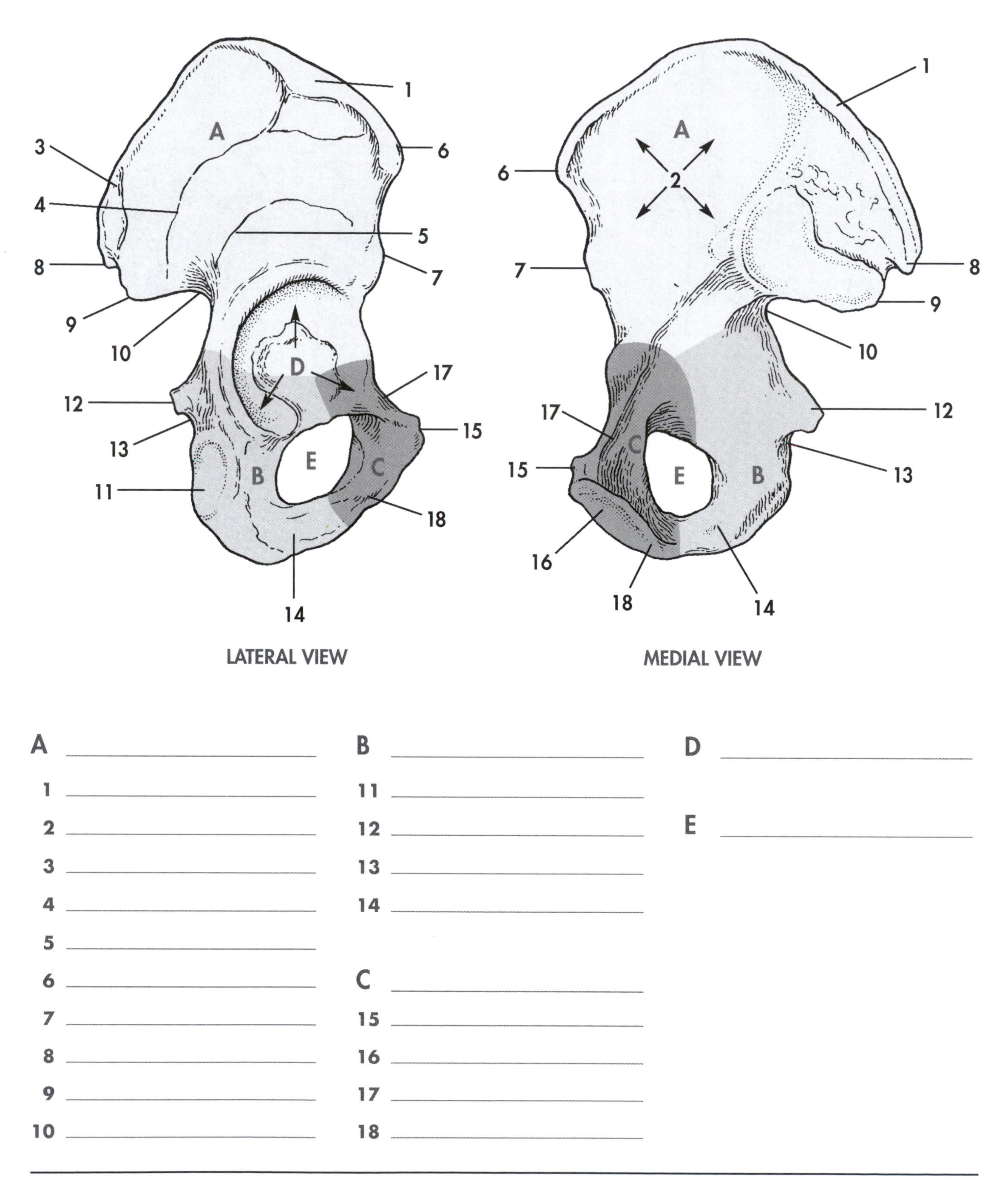

Right Femur

(fee•mur)

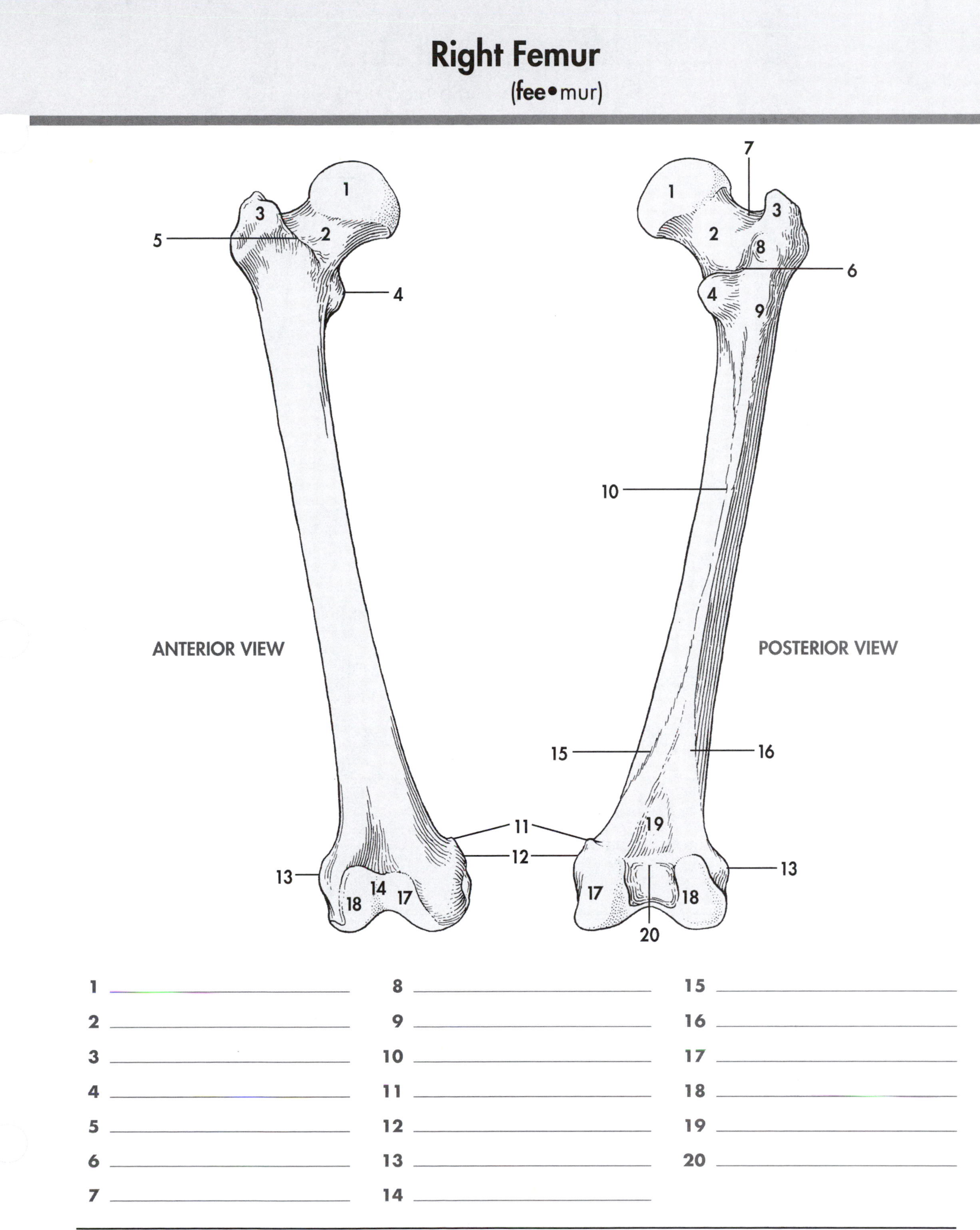

1 ____________ 8 ____________ 15 ____________

2 ____________ 9 ____________ 16 ____________

3 ____________ 10 ____________ 17 ____________

4 ____________ 11 ____________ 18 ____________

5 ____________ 12 ____________ 19 ____________

6 ____________ 13 ____________ 20 ____________

7 ____________ 14 ____________

Right Tibia and Fibula

(**tib**•ee•ah) (**fib**•yoo•lah)

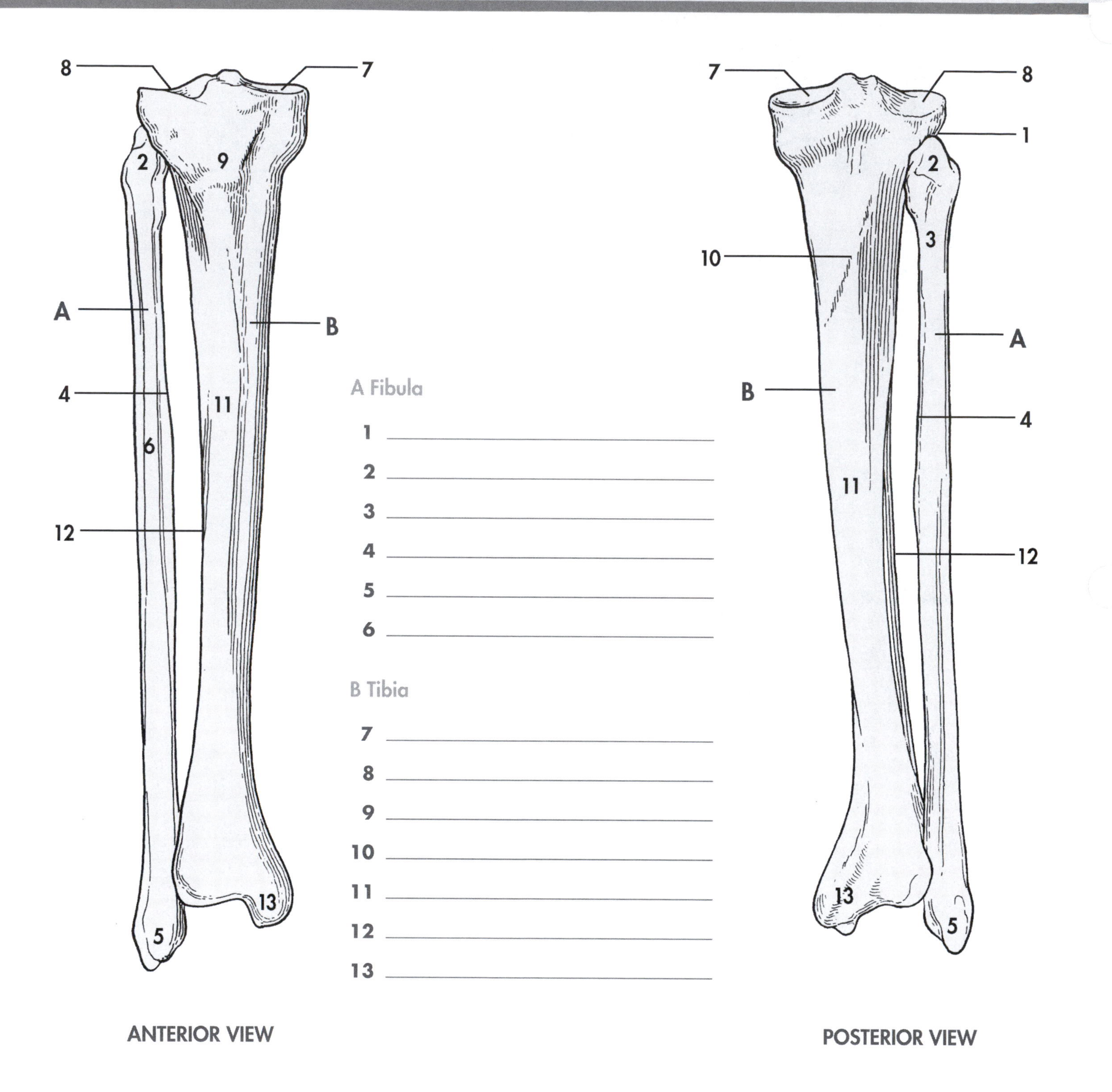

Right Foot and Ankle

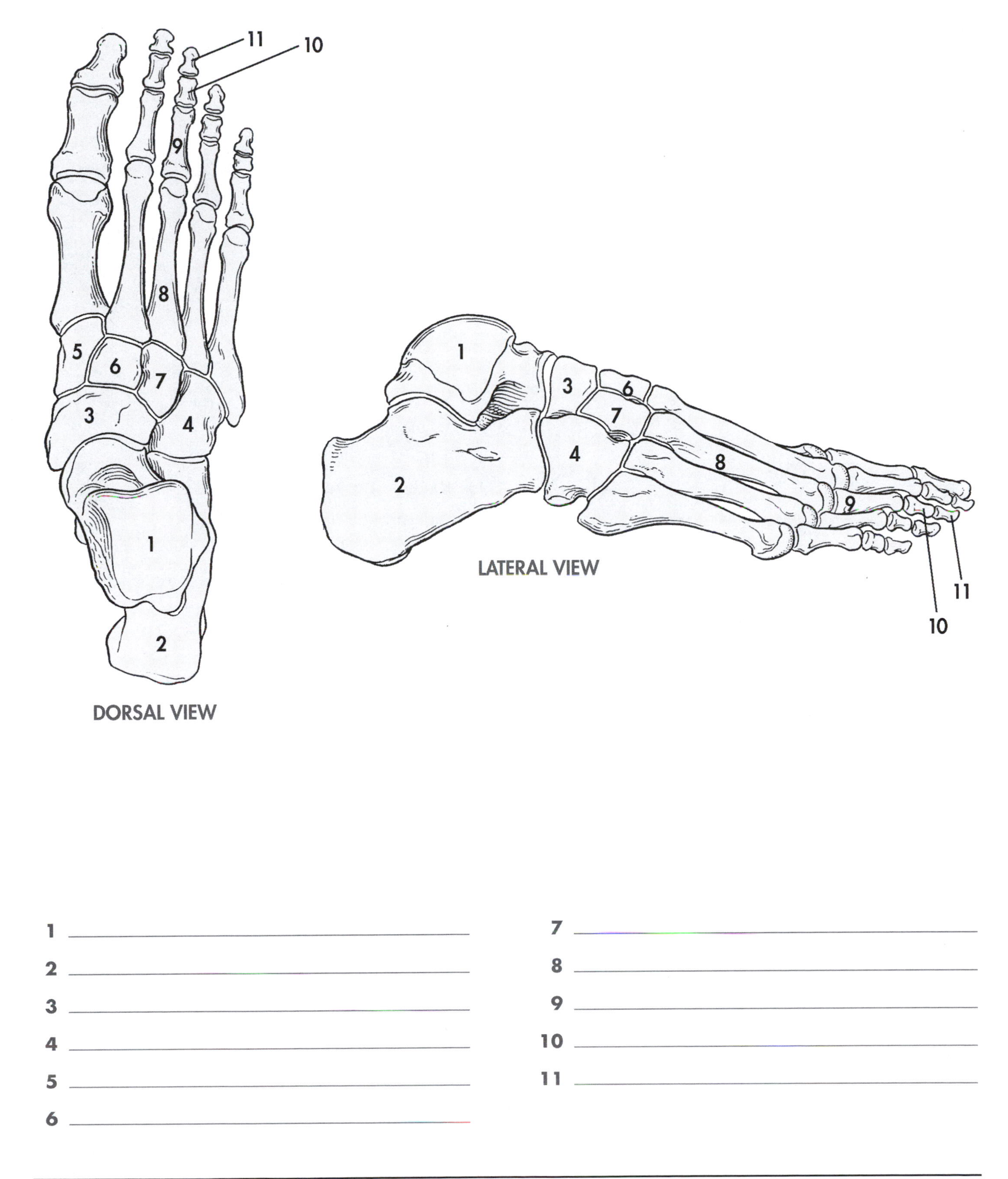

1 ______________________

2 ______________________

3 ______________________

4 ______________________

5 ______________________

6 ______________________

7 ______________________

8 ______________________

9 ______________________

10 ______________________

11 ______________________

II. MATCHING

Group A: In the blank to the left of each numbered bony landmark, write the **LETTER** of the bone on which it is located. Some lettered items will not be used.

BONY LANDMARKS

1. _____ Greater tubercle
2. _____ Olecranon process
3. _____ Lateral malleolus
4. _____ Zygomatic process
5. _____ Acromion process
6. _____ Xiphoid process
7. _____ Deltoid tuberosity
8. _____ Acetabulum
9. _____ Conoid tubercle
10. _____ Greater wing

CHOICES

A. Temporal Bone
B. Humerus
C. Sternum
D. Sphenoid
E. Os coxa
F. Ulna
G. Scapula
H. Femur
I. Clavicle
J. Fibula
K. Tibia
L. Radius
M. Calcaneus

Group B: The listed bony landmarks to the left are each found on more than one bone. In the blank to the left of each bony landmark, write the **LETTERS** of all the bones on which each type bony landmark is located. Each lettered bone might be used once, more than once, or not at all.

BONY LANDMARKS

1. _____ Crest
2. _____ Tuberosity
3. _____ Spine
4. _____ Head
5. _____ Coronoid process
6. _____ Styloid process

CHOICES

A. Tibia
B. Ilium
C. Scapula
D. Mandible
E. Ulna
F. Radius
G. Femur
H. Humerus
I. Vertebra
J. Temporal
K. Ischium
L. Pubis
M. Fibula

III. SENTENCE COMPLETION

Circle the term in parentheses that correctly completes each statement.

1. If kyphosis or "humpback" is an atypical posterior curvature of the vertebral column of the upper back, then kyphosis is an abnormality of the (**cervical, lumbar, sacral, thoracic**) vertebral region.
2. If you feel the "bumps" on either side at the distal end of the humerus, you would be palpating the (**condyles, epicondyles, malleoli, styloid processes**).
3. The type of fracture in which a bone is shattered into several pieces is called a (**comminuted, greenstick, spiral, stress**) fracture.
4. The tarsal bone that forms the ankle joint is the (**calcaneus, cuboid, navicular, talus**).
5. A hole in a bone for the passage of nerves and blood vessels is called a (**foramen, fossa, meatus, notch**).
6. In a typical long bone such as the humerus, the ends of the bone where you find the articular cartilages and various bony landmarks for muscle attachments are termed the (**diaphyses, epiphyses**).
7. The type of adolescent overuse injury/fracture in which the posterior growth plate partially or totally separates from the heel bone is known as a/an (**Osgood-Schlatter fracture, Sever's disease injury**).
8. Transverse foramina are found only in (**cervical, thoracic, lumbar, sacral**) vertebrae.
9. The humerus articulates with the (**acetabulum, glenoid fossa**).
10. When the hand is pronated, the (**radius swivels across the ulna, the ulna swivels across the radius**).
11. The (**clavicles and scapulae, os coxae**) make up the pectoral girdle.
12. There are (**5, 7, 12**) vertebrae in the cervical region of the vertebral column.
13. The large opening in the base of the skull that the spinal cord passes through is the foramen (**jugular, ovale, magnum, lacerum**).
14. The hard palate is made up of the palatine bones and the palatine portions of the (**zygomatic, maxilla, mandible, temporal**) bones.
15. During CPR, the part of the sternum that might break and puncture the heart or liver is the (**xiphoid process, manubrium, body, jugular notch**).
16. The most superior cervical vertebra is the (**axis, atlas**).
17. The carpal that articulates with the base of the thumb is the (**scaphoid, capitate, trapezium, pisiform**).
18. In the foot, the talus rests primarily on the calcaneous and articulates anteriorly primarily with the (**cuboid, navicular, first cuneiform, second cuneiform**).
19. In a/an (**avulsion, oblique, comminuted, stress**) fracture, a part of the bone is broken away.
20. A common site for an impacted fracture in the elderly is at the (**shoulder, hip, knee, elbow**) joint.

IV. MULTIPLE CHOICE

In the blank to the left of each question, write the **LETTER** of the correct answer.

1. _____ An abnormal sidewise curvature of the vertebral column is technically called
   A. kyphosis
   B. lordosis
   C. scoliosis

2. _____ The type of epiphyseal plate fracture in which the fracture continues from the growth plate through both the epiphysis and metaphysis is classified as a Salter-Harris Type _______ fracture.
   A. I
   B. II
   C. III
   D. IV
   E. V

3. _____ The distal end of the radius bone articulates with which two carpal bones?
   A. scaphoid and capitate
   B. hamate and capitate
   C. pisiform and triquetral
   D. scaphoid and lunate

4. _____ The odontoid process is found on the
   A. temporal bone
   B. humerus
   C. sacrum
   D. axis

5. _____ The midline depression at the top of the manubrium of the sternum is called the
   A. xiphoid notch
   B. jugular notch
   C. acromion fossa
   D. sternal angle

6. _____ The suture that separates the parietal bones from the occipital bone is the
   A. sagittal suture
   B. squamosal suture
   C. coronal suture
   D. lambdoidal suture

7. _____ The mastoid and styloid processes are located on the
   A. temporal bone
   B. sphenoid bone
   C. occipital bone
   D. parietal bone

8. _____ The crista galli, and the greater and lesser wings are all portions of the
   A. temporal bone
   B. sphenoid bone
   C. occipital bone
   D. parietal bone

9. _____ All of the following bones are part of the axial skeleton except the
   A. vertebrae
   B. scapula
   C. sternum
   D. occipital

10. _____ A roughened area of bone for the attachment of muscles is termed a
    A. condyle
    B. tuberosity
    C. fossa
    D. head

11. _____ The greater trochanter is located on the _______ bone.
    A. ulna
    B. humerus
    C. femur
    D. tibia

12. _____ The olecranon fossa is located on the _______ bone.

A. ulna
B. humerus
C. tibia
D. femur

13. _____ The hand bones are the _______ bones.

A. tarsal
B. metatarsal
C. carpal
D. metacarpal

14. _____ A type of fracture common in young gymnasts in which the epiphyseal plate is compressed is a Salter-Harris _________ fracture.

A. Type I
B. Type II
C. Type III
D. Type IV
E. Type V

15. _____ All of the following features are found on the femur **except** the

A. gluteal tuberosity
B. lateral epicondyle
C. tibial tuberosity
D. linea aspera

V. FILL IN THE BLANK

Write the correct word(s) in the blank spaces to complete each statement.

1. The three bones that "meet" in the shoulder are the ______________________, ______________________, and ______________________.

2. The suture between the two parietal bones is called the ______________________ suture.

3. A fracture, common in young children, in which one side of the bone breaks and the other side bends, is called a ______________________ fracture.

4. A fracture in which the bone protrudes to the outer surface of the skin is called a __________________fracture.

It is important to know both the "common" as well as the technical terms for parts of the skeleton, particularly when working with patients or clients who may only be familiar with the common terminology.

5. The "heel bone" is technically called the ______________________.

6. The zygomatic bone is commonly referred to as the ______________________.

7. The "knee cap" is technically called the ______________________.

8. The "funny bone," which is not a separate bone, is technically the _________________ process of the ulna.

9. The "collar bone" is technically called the ______________________.

10. Two technical terms for the lower jaw bone are the ______________________ and the ______________________.

11. The sternum is commonly referred to as the ______________________.

12. The "thumb" is technically called the ______________________ .

13. The os coxa is commonly referred to as the ______________________.

14. The scapula is commonly referred to as the ______________________.

15. The hallucis is commonly referred to as the ______________________.

VI. SHORT ANSWER

Answer each of the questions in the space provided.

1. Ribs can be divided into three groups on the basis of differences in how they articulate with the sternum. Briefly describe how each of the three groups of ribs articulate with the sternum and indicate the number of ribs in each group.

   A. "True" ribs: ______________________

   B. "False" ribs: ______________________

   C. "Floating" ribs: ______________________

2. In the development of the skeletal system, most bones initially develop as multiple "centers of ossification" that later fuse together to form single bones. The frontal bone and sternum each initially develop as two or more separate bones that later fuse into one bone. What is the functional advantage of the fused condition of each of these bones?

3. Write the correct spelling of each term in the blank.

   A. humorus ______________________

   B. vertabrae ______________________

   C. ocipital ______________________

   D. frackture ______________________

   E. radiel ______________________

   F. thorasic ______________________

4. What is a displaced fracture? Give three examples of displaced fractures.

5. What is the difference between an apophysis and an epiphysis?

VII. COMPARISON OF THE MALE AND FEMALE PELVIS

A. Examine the illustrations and label both pelvises with features 1–6 using the corresponding numbers (and arrows as necessary):

1 = Iliac Crest
2 = Sacrum
3 = Ischial Spine
4 = Pubic Angle
5 = Pelvic Inlet
6 = Coccyx

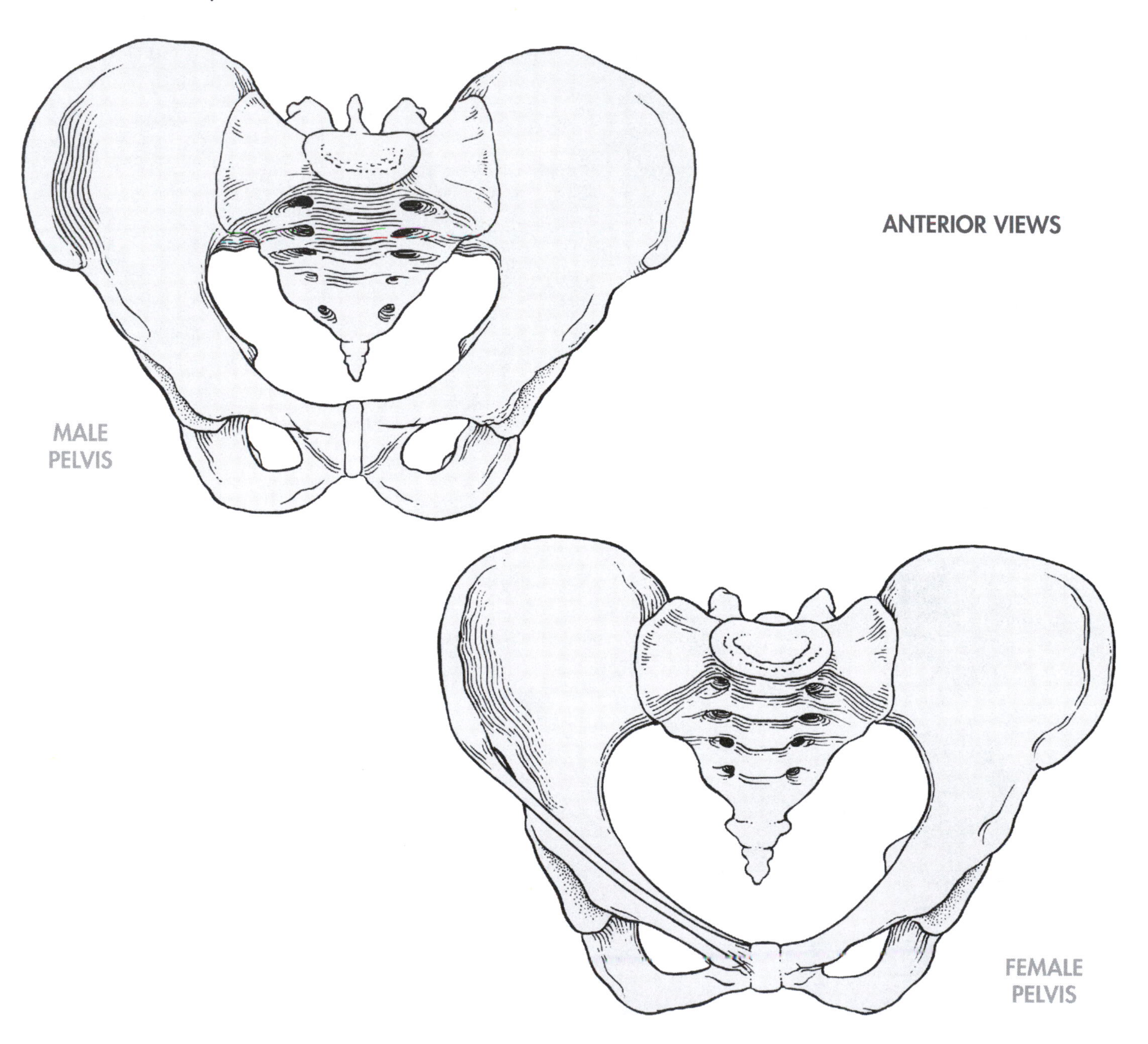

B. The female and male pelvises have the same general anatomy but vary in subtle details, such as size and shape of various features. Complete the following table by briefly indicating the specific condition of each feature in both the female and male pelvises to show the differences between the sexes.

LANDMARK	FEMALE	MALE
1. Iliac Crest		
2. Sacrum		
3. Ischial Spine		
4. Pubic Angle		
5. Pelvic Inlet		
6. Coccyx		

C. Briefly explain the significance of the anatomical differences between the female and male pelvises.

VIII. CASE STUDIES

A. A 12-year-old gymnast fell on an outstretched arm. She felt intense pain in the shoulder and couldn't move the arm. The shoulder sagged down and forward. There is a visible "lump" midway between her shoulder and sternum.

1. What type injury might be suspected? ______________________________

____

2. Which bone would be injured? ______________________________

____

3. Which muscle would pull up on the medial fragment of the injured bone? ______________________________

____

4. What causes the shoulder to sag? ______________________________

____

B. A football player received a blow to the lower right leg. He was in pain and not able to stand. X-rays showed a spiral fracture to the lateral bone.

1. What is the lateral bone of the lower leg? ______________________________

____

2. Briefly explain the characteristics of a spiral fracture. ______________________________

____

3. Which muscles overlie this bone? ______________________________

____

4. Which nerve might be damaged in this injury? ______________________________

____

5. Which foot actions might be impaired? ______________________________

____

IX. SELF/PARTNER PALPATION EXERCISES

If separate bones are available, one student puts on a blindfold and is given a bone. He/she should name and explain what part of the skeleton it is and why it is a right or left bone. Palpate and name as many bony landmarks as possible, describing the specific location of each in appropriate anatomical term(s) (distal, lateral, anterior-medial, etc.).

Now, without a blindfold, locate and palpate the bones and landmarks on yourself or a partner by following the written "road-map" directions.

SKULL

You will be able to palpate many of the skull bones.

- On the **dentary bone,** or **mandible,** feel along its lower margin, around the posterior angle, and upward toward the articulation with the temporal bone. To confirm the posterior margin, as you press along it, depress and elevate the lower jaw; you will feel the jaw rock back and forth as the **mandibular condyle** moves in the **temporal fossa.**
- Next, move your finger below and behind the lower edge of the earlobe and press upward against the blunt **mastoid process** of the **temporal bone.** Next, move your fingers in front of the external opening to the auditory canal and press your index finger against the horizontal, bony **zygomatic arch.** Press your index finger and thumb along the upper and lower edges of the zygomatic arch. Continue your index finger forward onto the prominent "cheek," or **zygomatic bone.** Next, move the index finger, down and medially across the **maxillary bone,** or **upper jaw,** to the base of the **nasal openings.**
- Next, place the three middle fingers of both hands on your "forehead" (**frontal bone**), move inferiorly to above the eyes and feel the raised bony "brow ridges" (will vary in prominence between males and females) beneath your eyebrows. Move your index fingers medially along the brow ridges and downward between the eyes onto the paired **nasal bones** that form the "bridge" of the nose. Feel anteriorly along the nasal bones until you reach the flexible, cartilaginous anterior portion of the nose. Flex this cartilage between your fingers and feel free to "wiggle your nose"! Return your index fingers to the lateral edge of each "brow ridge," and bring them inferiorly along the bar of bone on the lateral sides of the orbits and note that they continue onto the zygomatic "cheek" bone.
- Now palpate the other large bones of the skull that form the majority of the wall of the skull cavity, or **calvarium.** Place the thumb on the **mastoid process,** two fingers behind the earlobe and two fingers above the earlobe, to palpate the **temporal bone.** Move your four fingers superiorly onto the **parietal bone** on that side. Now shift your hand so that the palm of the hand "cups" the posterior–inferior portion of the skull, which is the **occipital bone.** The "bump" that you feel is the **external occipital protuberance.**

STERNUM

The component elements of the sternum are easily palpated.

- Place an index finger in the middle of the upper neck and run it down until you feel a shallow U-shaped notch. This is the **jugular notch,** the superior surface of the **manubrium,** the uppermost part of the sternum. Slightly to the side, while you rotate the shoulder, you can feel the movement of the medial end of the **clavicle,** or "collar bone," as it articulates with the manubrium. Move the finger slowly inferiorly and feel for a slight transverse ridge between the manubrium and the **body of the sternum.** Proceeding inferiorly, you will readily feel the narrower **xiphoid process.** The ribs articulate laterally with the sternum individually ("true ribs") or jointly by costal cartilages ("false ribs"). The last two ribs are shortened in length and do not articulate ("false floating ribs") with the sternum.

ULNA

- Hold the lower arm to be examined in a supine position. The **ulna** lies on the medial side of the lower arm. At the upper (proximal) end, palpate the prominent "funny bone" as the lower arm is flexed and extended. Note that the funny bone moves with the ulna, not separate from it; it is the **olecranon process** of the ulna. Feel along the sharp, medial surface of the shaft of the ulna to its lower end. Feel the slim, pointed **styloid process.** To confirm the process is part of the ulna and not one of the wrist bones, flex and extend the wrist and note that the styloid process does not move when the wrist moves. Press adjacent to this process to feel the **head of the ulna**.

TIBIA

- Place the fingers of one hand around the edge of the "knee cap," or **patella bone,** and actively move the lower leg to confirm that the patella articulates with both the femur and the tibia. Move two fingers from the inferior edge of the patella immediately onto the roughened, anterior–medial **tibial tuberosity** on which the quadriceps muscle inserts. Feel along the length, or "shaft," **of the tibia.** At the inferior end of the tibia, place your thumb on the "bump," or the **medial malleolus,** of the tibia (the lateral "bump" is the **lateral malleolus** of the fibula). The malleoli are commonly and mistakenly referred to as the "ankle bones" but are actually processes of the tibia and fibula that overlap the talus bone medially and laterally, stabilizing sidewise movement of the "real" ankle bone. Place a finger on each malleolus and dorsiflex and plantar flex the foot, and move it both medially and laterally to demonstrate this relationship.

ANKLE and FOOT

This will have to be done with a partner.

- Use the medial and lateral malleoli that you located and palpated as reference landmarks to pursue ankle and foot bone palpations. Move your fingers from the malleoli down to the medial and lateral sides of the "heel bone," the **calcaneus,** and feel around its contours. The ankle joint is primarily the articulation between the **talus bone** and the distal tibia; the medial and lateral malleoli act as sidewise stabilizers. Hold the subject's right foot "limp" with your right hand, and plantar flex and invert the foot. This will enable your left thumb to palpate toward the anterior part of the convex surface of the **talar "dome"** of the talus that articulates with the tibia.
- Now move on to identify and palpate the remaining five **tarsal bones.** Hold the foot across the arch with your left hand. The elongate **navicular** extends transversely from the medial edge of the foot between the cuneiforms anteriorly, and the talus posteriorly. Lay the thumb of your right hand across the length of the navicular, feeling for the crevices in front and behind it to make sure you are on the navicular. Now slide your thumb distally to lay your thumb across the **medial, middle,** and **lateral cuneiforms.** Reach the thumb further laterally, and you will be palpating the **cuboid** that lies lateral to the navicular and the lateral cuneiform.
- Your left hand has been holding the foot across the **metatarsals.** Slide that hand posteriorly to hold the foot under the **tarsals.** With the thumb of the right hand, feel along the length of each of the five metatarsals. Note the shape of both ends of the metatarsals and especially the posterior lateral projection of the fifth metatarsal. Now palpate the **phalanges** of the toes. There are only two phalanges in the **"big toe,"** or **hallucis,** and three in each of the other four toes. The articulations between the metatarsals and proximal phalanges are condyloid but not as moveable sidewise as in the hand. The interphalangeal articulations are hinge joints as in the hand.

Articulations and Body Motions

2

At the end of this chapter, you should be able to

1. Differentiate between tendons and ligaments.
2. Give the function of a bursa.
3. Differentiate among fibrous, cartilaginous, and synovial joints.
4. Give three examples of fibrous joints.
5. Give two examples of cartilaginous joints.
6. Give examples of the six types of synovial joints.
7. Explain the type of movement allowed by fibrous joints.
8. Explain the type of movement allowed by cartilaginous joints.
9. Explain the type of movement allowed by the six types of synovial joints.
10. Identify the components of a typical synovial joint.
11. Identify the components of the following joints: shoulder, elbow, hip, knee.
12. Compare the stability and range of motion in the shoulder and hip joint.
13. Define the following terms: flexion, extension, abduction, adduction, rotation, circumduction, opposition, pronation, supination, inversion, eversion, dorsiflexion, plantar flexion.
14. Give an example of a joint in which each of the preceding motions occur.

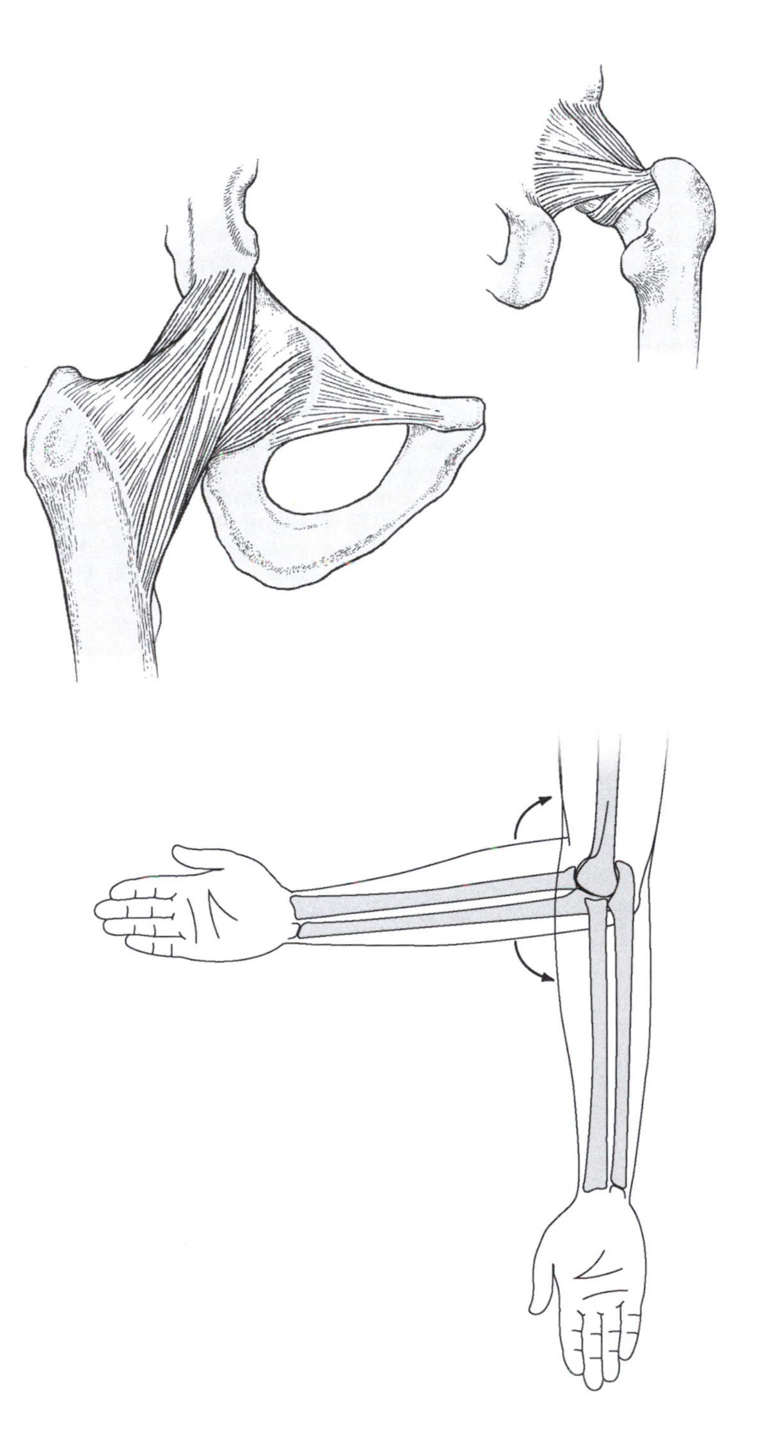

Word Search Puzzle

INSTRUCTIONS: Find and CIRCLE each of the listed terms in the Word Search Puzzle.
(Terms may read from left to right, right to left, up, down, or diagonally.)

S Y N O V I A L Z A E V X Z V W S X H
Y X Z P Q Q X A Q L T Q X W Q U N I Y
N Z X S U T U R E L A P Z Z C X N W P
D E P R E S S E V E V X X S W G I X E
E Q V S E R E T W T E V I A E F X Z R
S A D D L E Z A X A L N F D K D E W E
M W V Q R X F L K P E W R D Q I L V X
O Z X V W M N L Q M D Q O U Z O F Q T
S I S O H P M O G P B Z T C Z L I Q E
I P Q P I M N C R U C I A T E Y S P N
S W X V X N F X Z C F P T I V D R P S
Z X O W Z W G B F L E X I O N N O B I
W T K S Y M P H Y S I S O N X O D B O
N I O T I S O P P O Z W N X W C C C N

ADDUCTION
COLLATERAL
CONDYLOID
CRUCIATE
DEPRESS
DORSIFLEX
ELEVATE
FLEXION
GOMPHOSIS
HINGE

MENISCUS
OPPOSITION
PIVOT
ROTATION
SADDLE
SYMPHYSIS
SYNDESMOSIS
SYNOVIAL
SUTURE
TERES

I. BODY MOTION IDENTIFICATION

Write the technical terms for the **LETTERED** body motions in the lettered spaces beneath the illustrations.

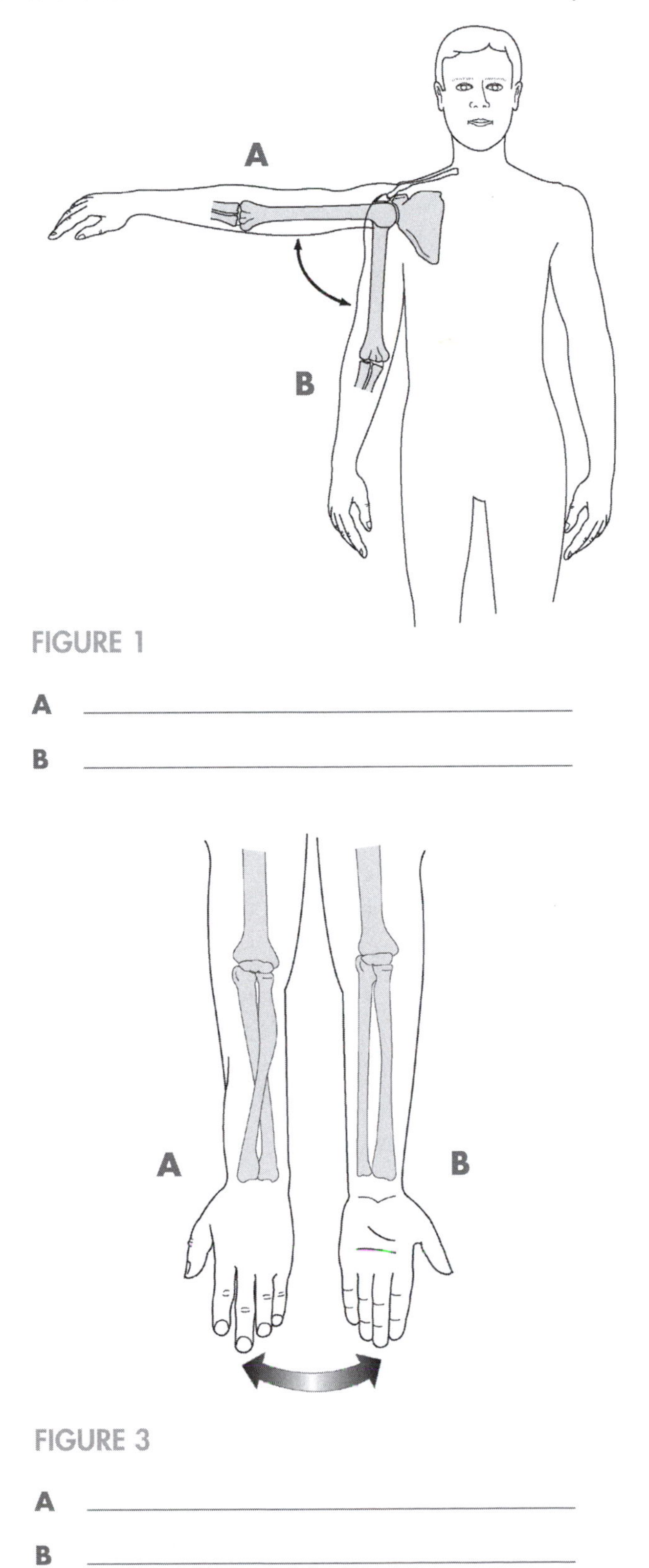

FIGURE 1

A ______________________________

B ______________________________

FIGURE 3

A ______________________________

B ______________________________

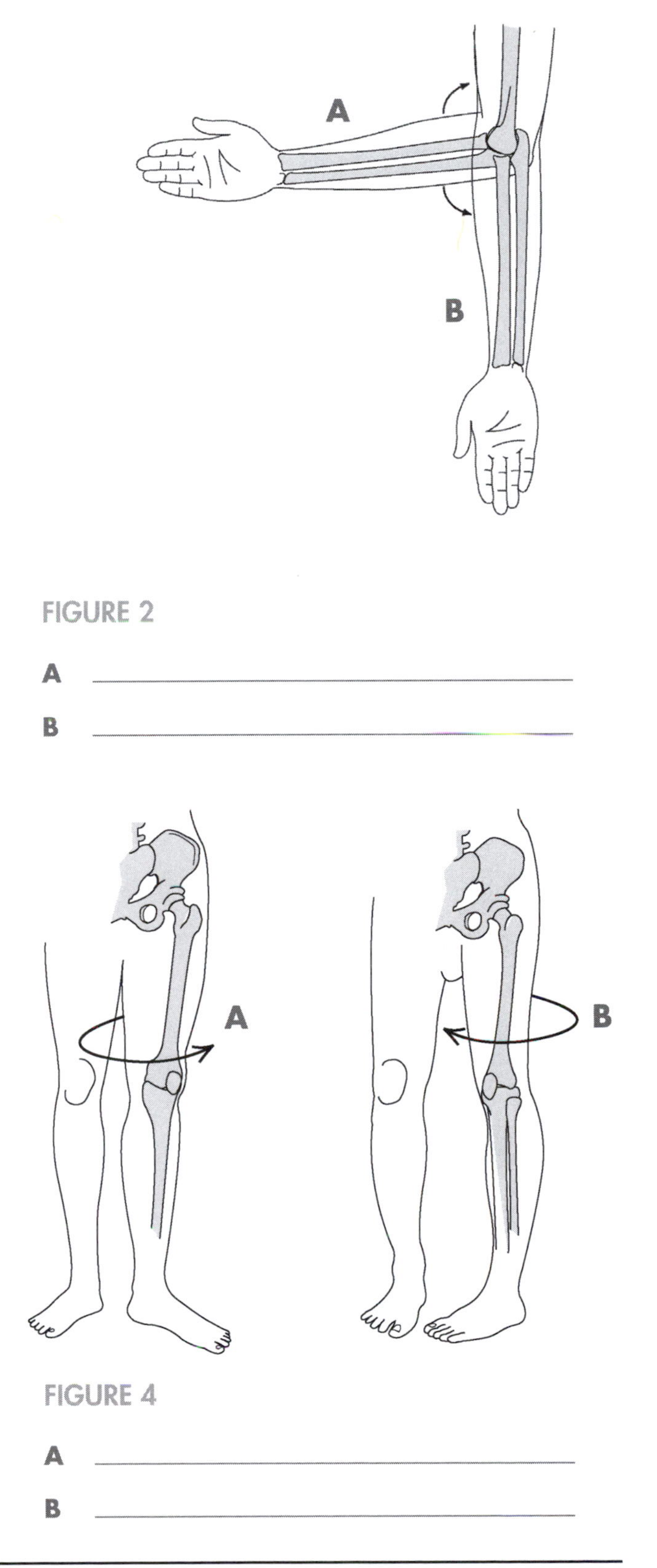

FIGURE 2

A ______________________________

B ______________________________

FIGURE 4

A ______________________________

B ______________________________

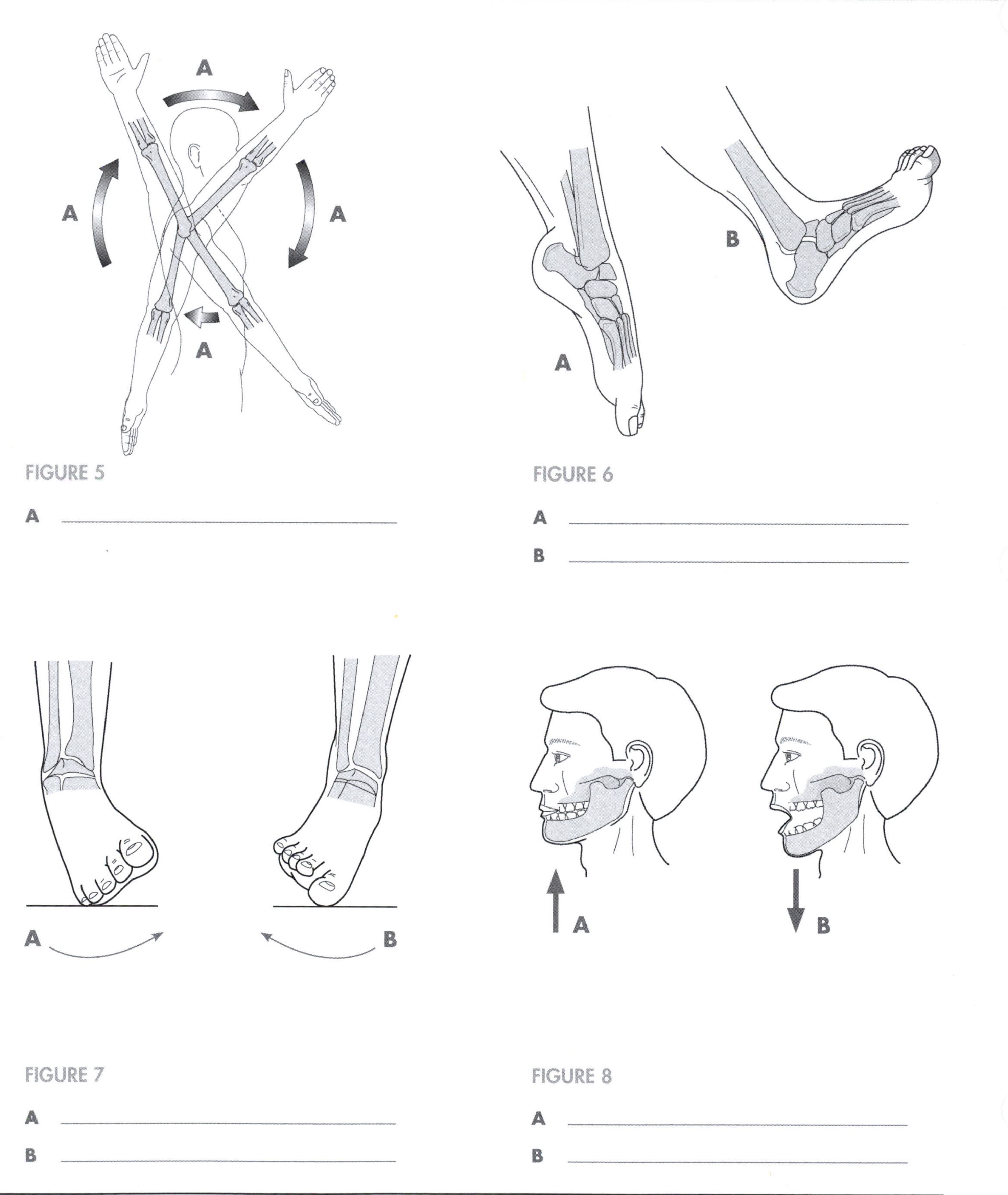

FIGURE 5

A ______________________

FIGURE 6

A ______________________

B ______________________

FIGURE 7

A ______________________

B ______________________

FIGURE 8

A ______________________

B ______________________

II. MATCHING

Group A: In the blank to the left of each statement, write the **LETTER** of the term that describes that articulation. Each lettered choice may be used once, more than once, or not at all.

1. _____ First metacarpal/carpal articulation
2. _____ Articulation permitting one to indicate "no" with his or her head
3. _____ Metacarpal/phalange articulation with digits 2–5
4. _____ Femoral/os coxa (coxal) articulation
5. _____ Articulation type permitting only uniaxial movement

A. Ball and socket
B. Condyloid
C. Gliding
D. Hinge
E. Pivot
F. Saddle

Group B: In the blank to the left of each statement, write the LETTER(s) of the types of articulations that exhibit the motions described. Each lettered choice may be used once or more than once.

1. _____ Permits rotation
2. _____ Permits flexion and extension
3. _____ Permits circumduction
4. _____ Permits opposition
5. _____ Permits abduction and adduction
6. _____ Permits pronation and supination

A. Ball and socket
B. Condyloid
C. Gliding
D. Hinge
E. Pivot
F. Saddle

Group C: In the blank to the left of each statement, write the **LETTER** of the type of articulation described by that statement.

1. _____ Between tooth and tooth socket in jaw bone
2. _____ Between adjacent vertebral bodies
3. _____ Between humerus and scapula
4. _____ Between carpal bones
5. _____ Between odontoid process of the axis and the atlas bone
6. _____ Between the humerus and ulna
7. _____ Between the distal ends of metacarpals and proximal ends of phalanges 2–5
8. _____ Between the base of the thumb and trapezium
9. _____ Between the distal ends of the tibia and fibula
10. _____ Between the bones of the cranium

A. Ball and socket
B. Condyloid
C. Gliding
D. Hinge
E. Pivot
F. Saddle
G. Suture
H. Symphysis
I. Gomphosis
J. Syndesmosis

III. SENTENCE COMPLETION

Circle the term in parentheses that correctly completes each statement.

1. All of the following are fibrous joints *except* a (**gomphosis, symphysis, syndesmosis, suture**).
2. The type of fibrous joint that allows essentially no movement is a (**suture, syndesmosis, symphysis, synchondrosis**) joint.
3. A type of joint with matching concave and convex surfaces that allows movement in two directions is a (**pivot, condyloid, hinge, gliding**) joint.
4. A type of joint with flat surfaces found between the tarsal bones is a (**condyloid, hinge, gliding, saddle**) joint.
5. The ligament that holds the head of the radius adjacent to the ulna notch, allowing the radius to pivot, is the (**annular, radial collateral, anterior cruciate, ligamentum teres**) ligament.
6. The type of joint between the acromion process of the scapula and the acromial end of the clavicle is a (**hinge, gliding, ball and socket, condyloid**) joint.
7. The joint between the head of the radius and the radial notch of the ulna is a (**pivot, saddle, ball and socket, hinge**) joint.
8. The type of joint between the two pubic bones is a (**gliding, symphysis, syndesmosis, gomphosis**) joint.
9. The interphalangeal joints are (**condyloid, saddle, gliding, hinge**) joints.
10. The joint between the head of the femur and the acetabular fossa of the os coxa is a (**hinge, ball and socket, gliding, pivot**) joint.
11. In the elbow, the ligament that connects the humerus and radius is the (**radial collateral, ulna collateral, annular**) ligament.
12. In the hip, the strong ligament that attaches to the anterior inferior iliac spine, the acetabular rim, and the femur is the (**ischiofemoral, pubofemoral, iliofemoral**) ligament.
13. Movement toward the midline is (**abduction, adduction**).
14. Movement of the arm to cause the hand to face forward is (**pronation, supination**).
15. Movement of the foot that causes the toes to point down is (**plantar flexion, dorsiflexion**).

IV. MULTIPLE CHOICE

In the blank to the left of each question, write the LETTER of the correct answer.

1. _____ Which one of the following types of joints is often temporary?
   A. Synovial
   B. Synchondrosis
   C. Fibrous
   D. Symphysis

2. _____ The joint between the occipital condyles and the atlas is a _____ joint.
   A. Pivot
   B. Condyloid
   C. Hinge
   D. Symphysis

3. _____ The joint between the navicular and cuboid is an example of a _____ joint.
   A. Gliding
   B. Symphysis
   C. Hinge
   D. Condyloid

4. _____ In the hip, the cup-shaped recess into which the head of the femur fits is called the
   A. Glenoid fossa
   B. Acetabular fossa
   C. Iliac fossa
   D. Obturator fossa

5. ______ The joints between the distal end of the metatarsals and the proximal end of phalanges 2 to 5 are ______ joints.
   A. Gliding
   B. Pivot
   C. Hinge
   D. Condyloid

6. ______ The type of movement allowed at a pivot joint is
   A. Flexion
   B. Adduction
   C. Rotation
   D. Circumduction

7. ______ The turning of the sole of the foot outward is an example of
   A. Pronation
   B. Supination
   C. Inversion
   D. Eversion

8. ______ A joint that contains a flat disk of fibrocartilage such as the one between the bodies of the vertebrae is a ______ joint.
   A. Synovial
   B. Synchondrosis
   C. Symphysis
   D. Syndesmosis

9. ______ The ______ ligament provides stability on the tibial side of the knee joint.
   A. Anterior cruciate
   B. Posterior cruciate
   C. Medial collateral
   D. Lateral collateral

10. ______ The type of joint that allows the greatest degree of movement is the ______ joint.
   A. Hinge
   B. Condyloid
   C. Gliding
   D. Ball and socket

11. ______ The joints found between the facets of adjacent vertebrae are ______ joints.
   A. Hinge
   B. Condyloid
   C. Gliding
   D. Symphysis

12. ______ Articular cartilages are only found in ______ joints.
   A. Symphysis
   B. Syndesmoses
   C. Synovial
   D. Synchondroses

13. ______ All of the following are condyloid joints *except*
   A. Radius–carpal
   B. Occipital condyle–atlas
   C. Carpometacarpal of thumb
   D. Temporomandibular

14. ______ A joint that allows movement in two directions is considered a ______ joint.
   A. Uniaxial
   B. Biaxial
   C. Multiaxial
   D. Triaxial

15. ______ A gliding joint is a ______ joint.
   A. Uniaxial
   B. Biaxial
   C. Multiaxial
   D. Nonaxial

V. FILL IN THE BLANK

Write the correct word(s) in the blank spaces to complete each statement.

1. The three types of fibrous joints are _______________, _______________, and _______________.
2. The type of cartilaginous joint found in the epiphyseal plate of long bones is a _______________.
3. The type of joint that has a fluid-filled joint cavity is a _______________ joint.
4. The temporomandibular joint (TMJ) is a _______________ joint.
5. The only saddle joint is found between the _______________ of the wrist and the _______________ at the base of the thumb.
6. The joint between the odontoid process of the axis and the atlas is a _______________ joint.
7. When the gastrocnemius contracts and elevates the calcaneus and other tarsal bones, the foot undergoes _______________.
8. Thrusting the jaw forward is an example of _______________.
9. The ligaments that give front to back stability in the knee are the _______________ and _______________ ligaments.
10. The ligament that attaches the head of the femur to the fovea in the acetabulum is the _______________.
11. The _______________, _______________, _______________, and _______________ ligaments form the lengthwise arch of the foot.
12. Plantar fasciitis occurs when the _______________ becomes inflamed.
13. A motion in which a bone turns around its own longitudinal axis is _______________.
14. Decreasing the angle of a hinge joint results in _______________.
15. Movement of the leg away from the midline is _______________.

VI. SHORT ANSWER

Answer each of the questions in the space provided.

1. Compare the stability and range of motion of the shoulder joint and the hip joint. Explain the reasons for the differences.

2. The temporomandibular joint previously was considered a hinge joint. Now it is more appropriately considered a condyloid joint. Why is the current classification more accurate for this joint?

3. What is the function of the menisci in the knee joint?

4. What is the function of a bursa around a joint?

5. Name two functions of the malleoli in the ankle joint.

VII. CASE STUDIES

A. A soccer player was hit on the posterolateral side of the left knee while his foot was firmly planted on the ground. His lower leg was sharply abducted by the impact.

1. Which ligaments were probably injured?

2. Which other structure in the knee might have been injured?

3. What are two roles of the cruciate ligaments?

4. What is the function of the menisci?

B. A 70-year-old patient presented with acute lower back pain that developed when he stood erect after picking a heavy package off a chair. The patient noted that he had had a similar incident six months earlier, with periodic pain in the buttock, back of the upper leg, but not the lower leg. He added that he had discomfort leaning back but not leaning forward. On palpation, the doctor noted that the pain was on the right side of the spine adjacent to lumbar vertebrae L4–L5. The doctor also noticed that the patient held his back in a stiff, out-of-alignment position.

1. What are the two types of articulations through which lumbar vertebrae articulate with each another?

2. Which intervertebral problems could develop with each type of articulation?

3. Which related neuromuscular conditions might develop from these vertebral articulation problems?

4. On the basis of the symptoms the patient related and the results of the doctor's visual and palpation examination, with which vertebral column problem does it seem that the patient is afflicted? Why?

Muscles of the Head, Neck, and Torso

3

At the end of this chapter, you should be able to

1. Identify the major muscles of the head and face.
2. Describe the actions of the muscles of the head and face.
3. Identify the neck muscles and give their attachments.
4. Describe the action of each of the neck muscles.
5. Identify the major trunk muscles and give their attachments.
6. Describe the action of each of the trunk muscles.
7. Identify the muscles used in inspiration.
8. Identify the muscles used in forced expiration.
9. Identify the cranial nerves innervating the facial muscles.
10. Define the terms "synergist" and "antagonist."
11. Explain the specific function of each of the six eye muscles.
12. Explain the cause of Bell's palsy and list the symptoms.

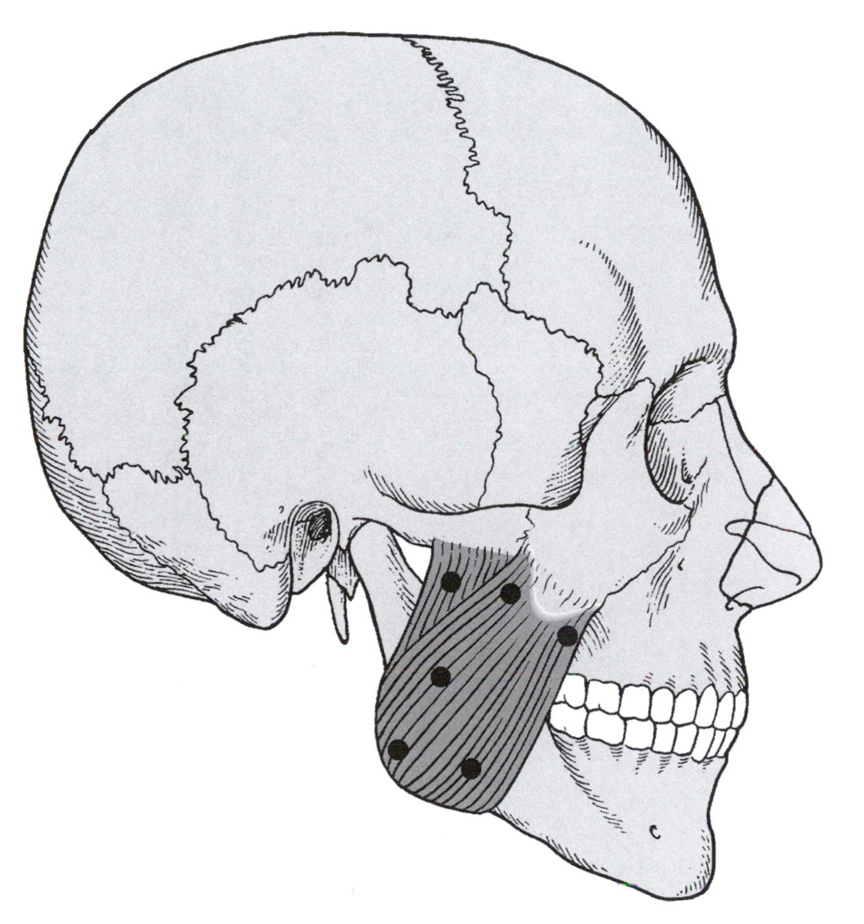

Word Search Puzzle

INSTRUCTIONS: Find and CIRCLE each of the listed terms in the Word Search Puzzle.
(Terms may read from left to right, right to left, up, down, or diagonally.)

R I S O R I U S X Z R O T A V E L W X Z E
E P P M X F R O N T A L I S Z W X B K C U
C X L Y Z W E V X Y Z P S U T A R D A U Q
T B E L N Z T Z X C A P I T I S P R Q Q I
U C N O A M E W Z Q Z S R Z P S U G N O L
S W I H S B S P V Y I O R P W V Z A X Q B
A X U Y A Q S V G D T C S W X Y Z M Z R O
B Q S O L W A O I A B U M C B X X S Y O L
D Q W I I Z M F N Z T D I O H O L Y M T A
O P W D S A I I F A G G X W X Q Z T Z C N
M Q V W T T C Q R O T A T O R E S A Z E R
I F B I L C W R D I A P H R A G M L Q R E
N Z C U U Y E P X Z Q S I L A R O P M E T
I U M B N S P I N A L I S P Q S W M Q Z X
S T E R N O H Y O I D F Y Z X S P I N A E

BUCCINATOR
CAPITIS
DIAPHRAGM
ERECTOR
EXTERNAL OBLIQUE
FRONTALIS
LEVATOR
LONGUS
MASSETER
MULTIFIDUS
MYLOHYOID
NASALIS
PLATYSMA
QUADRATUS
RECTUS ABDOMINIS
RISORIUS
ROTATORES
SERRATUS
SPINAE
SPINALIS
SPLENIUS
STERNOHYOID
TEMPORALIS

I. ILLUSTRATION IDENTIFICATION (pages 47–60 can be cut and used as flash cards)

Write the technical information for each muscle in the spaces beneath the illustrations.

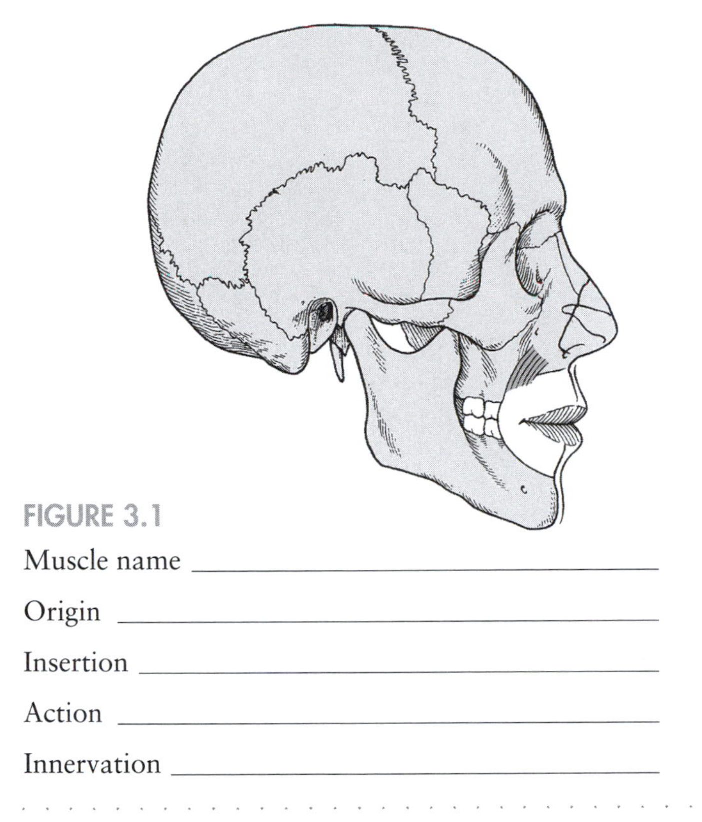

FIGURE 3.1

Muscle name ______________________

Origin ______________________

Insertion ______________________

Action ______________________

Innervation ______________________

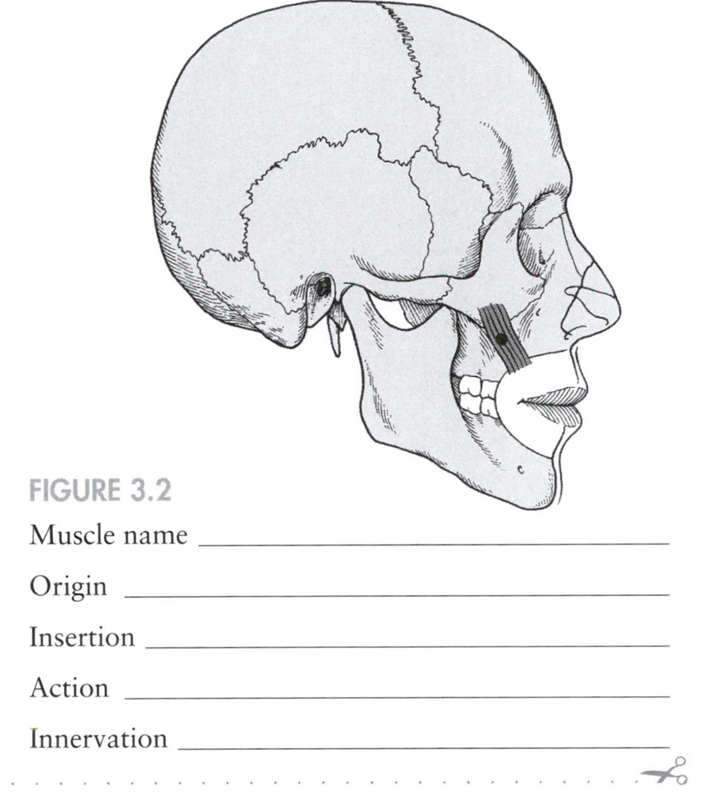

FIGURE 3.2

Muscle name ______________________

Origin ______________________

Insertion ______________________

Action ______________________

Innervation ______________________

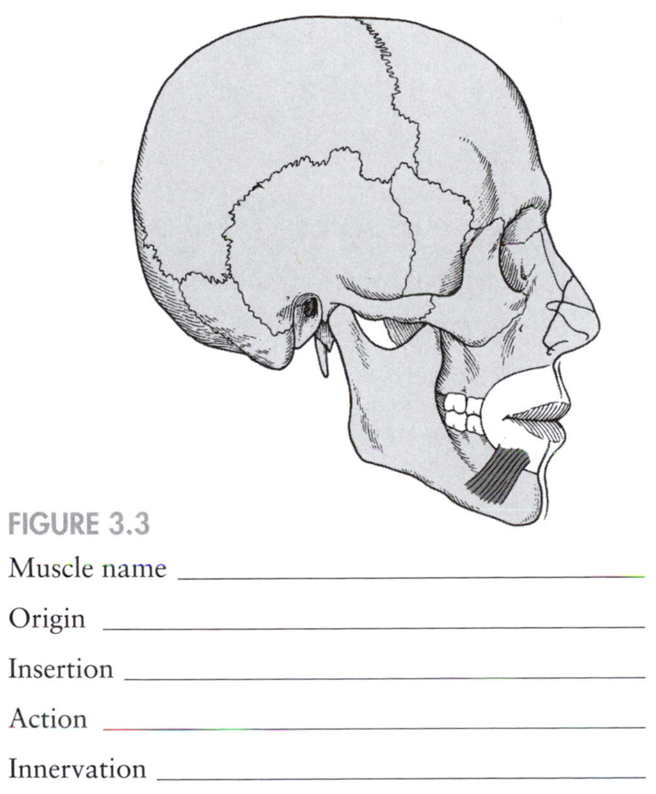

FIGURE 3.3

Muscle name ______________________

Origin ______________________

Insertion ______________________

Action ______________________

Innervation ______________________

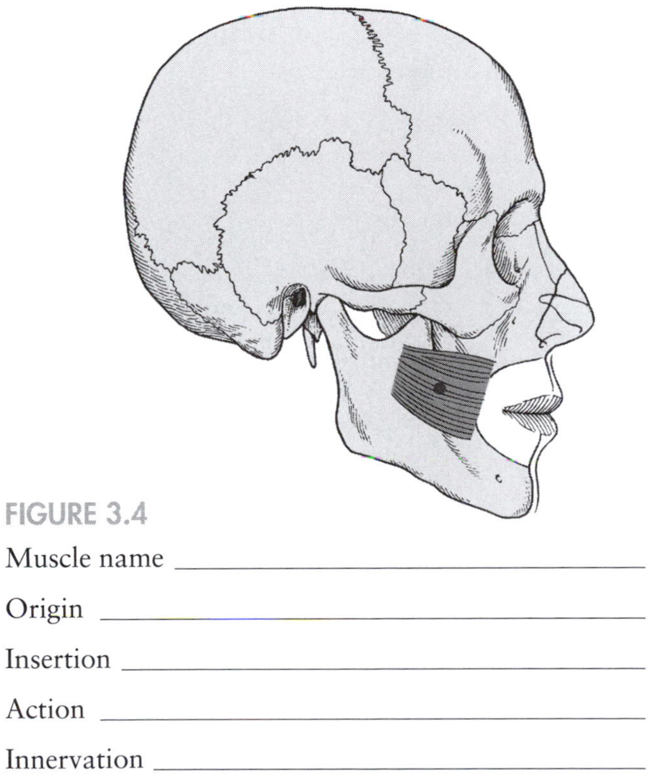

FIGURE 3.4

Muscle name ______________________

Origin ______________________

Insertion ______________________

Action ______________________

Innervation ______________________

FIGURE 3.2

Muscle name	Zygomaticus Major—Zygomatic Major
Origin	Zygomatic bone
Insertion	Angle of mouth blending with the levator anguli oris, orbicularis oris, and depressor anguli oris muscles
Action	Draws angle of mouth upward and outward
Innervation	Buccal branch of facial nerve (VII)

FIGURE 3.1

Muscle name	Levator Anguli Oris
Origin	Canine fossa of maxilla
Insertion	Angle of mouth, blending with fibers of zygomaticus major, depressor anguli oris and orbicularis oris muscles
Action	Elevates corners of mouth
Innervation	Buccal branch of facial nerve (VII)

FIGURE 3.4

Muscle name	Buccinator
Origin	Outer surface of alveolar processes of maxilla and mandible and pterygomandibular raphe
Insertion	Angle of mouth blending with fibers of the orbicularis oris muscle
Action	Draws corner of mouth laterally, compresses cheek
Innervation	Lower buccal branches of the facial nerve (VII)

FIGURE 3.3

Muscle name	Depressor Labii Inferioris
Origin	Body of the mandible lateral to the midline, between the mandibular symphysis and the mental foramen
Insertion	Skin and muscle of lower lip blending with fibers of orbicularis oris
Action	Draws lower lip inferiorly and laterally during mastication
Innervation	Mandibular branch of facial nerve (VII)

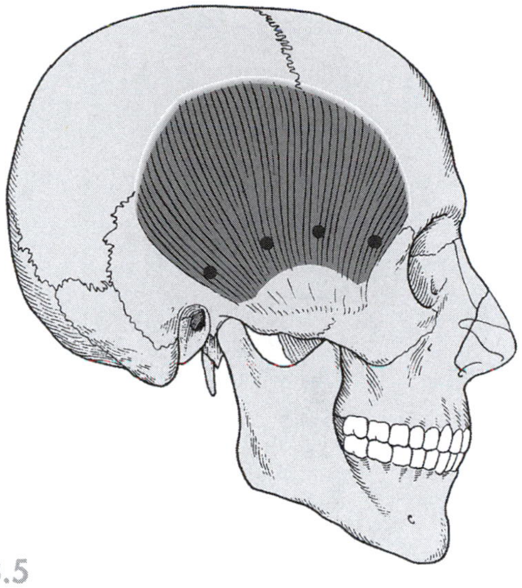

FIGURE 3.5

Muscle name ______

Origin ______

Insertion ______

Action ______

Innervation ______

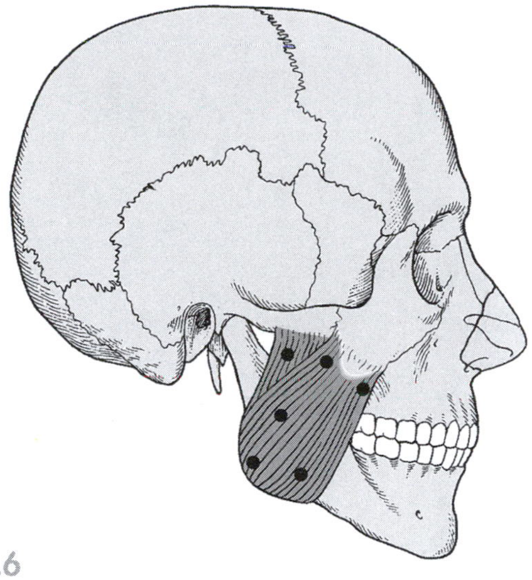

FIGURE 3.6

Muscle name ______

Origin ______

Insertion ______

Action ______

Innervation ______

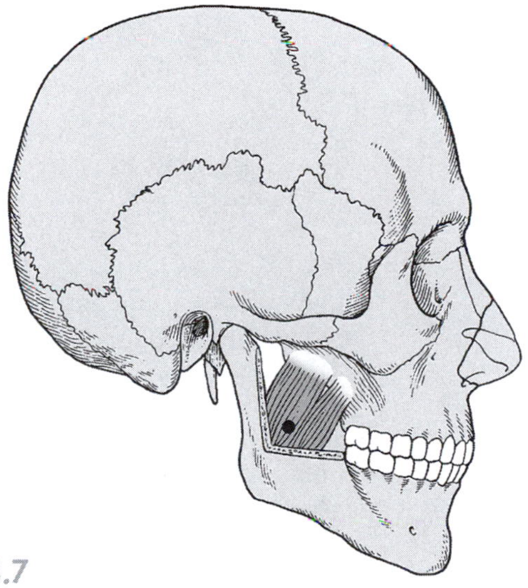

FIGURE 3.7

Muscle name ______

Origin ______

Insertion ______

Action ______

Innervation ______

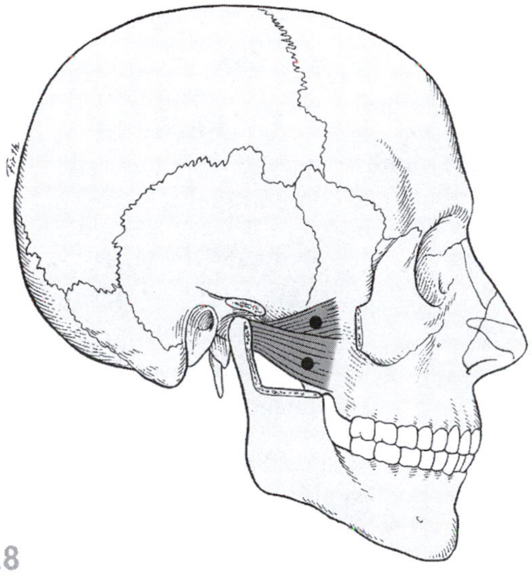

FIGURE 3.8

Muscle name ______

Origin ______

Insertion ______

Action ______

Innervation ______

FIGURE 3.6

Muscle name	Masseter
Origin	Zygomatic process of maxilla and medial and inferior surfaces of zygomatic arch
Insertion	Angle and ramus of the mandible
Action	Elevates mandible and slightly protracts it
Innervation	Mandibular branch of trigeminal nerve (V)

FIGURE 3.5

Muscle name	Temporalis
Origin	Temporal fossa and temporal fascia
Insertion	Coronoid process of mandible via a tendon that passes deep to the zygomatic arch
Action	Elevates and retracts mandible, assists in side-to-side movement of mandible
Innervation	Mandibular branch of trigeminal nerve (V)

FIGURE 3.8

Muscle name	Pterygoideus Lateralis—Lateral Pterygoid
Origin	A superior head arises from the greater wing of sphenoid bone; an inferior head from the lateral surface of the lateral pterygoid plate of sphenoid
Insertion	Both heads insert on the mandibular condyle and temporomandibular joint capsule
Action	Protrudes, depresses, and moves mandible from side-to-side
Innervation	Mandibular branch of trigeminal nerve (V)

FIGURE 3.7

Muscle name	Pterygoideus Medialis—Medial Pterygoid
Origin	Medial surface of lateral pterygoid plate of sphenoid, and the maxilla and palatine bones
Insertion	Posteroinferior aspect of the medial surface of ramus and angle of the mandible
Action	Synergistic with temporalis and masseter in elevation of mandible; it causes protrusion and side-to-side movements of the mandible
Innervation	Mandibular branch of trigeminal nerve (V)

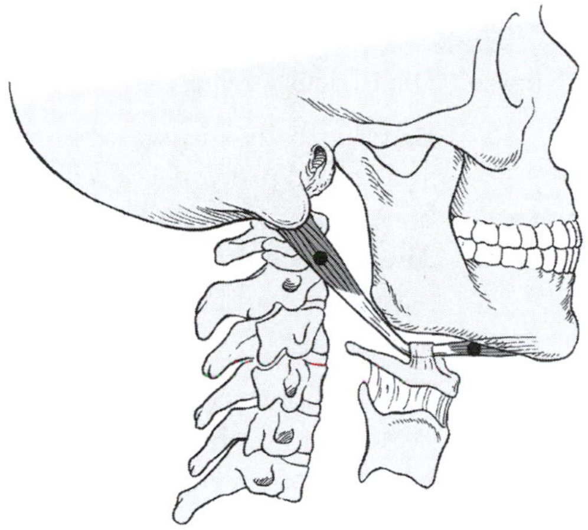

FIGURE 3.9

Muscle name ______

Origin ______

Insertion ______

Action ______

Innervation ______

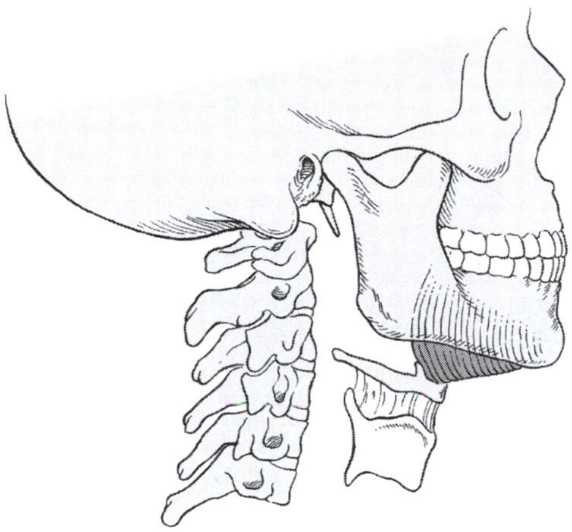

FIGURE 3.10

Muscle name ______

Origin ______

Insertion ______

Action ______

Innervation ______

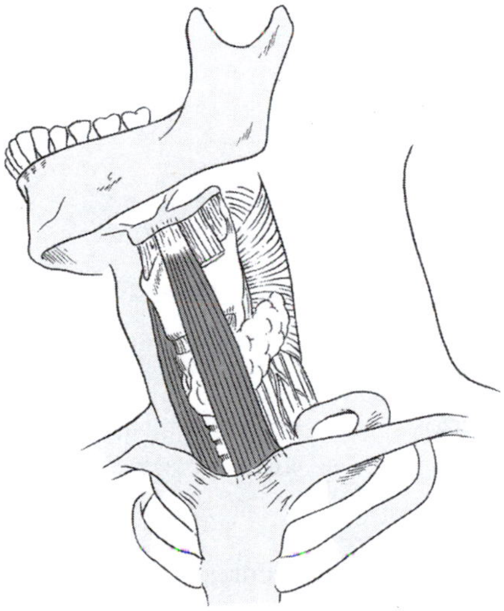

FIGURE 3.11

Muscle name ______

Origin ______

Insertion ______

Action ______

Innervation ______

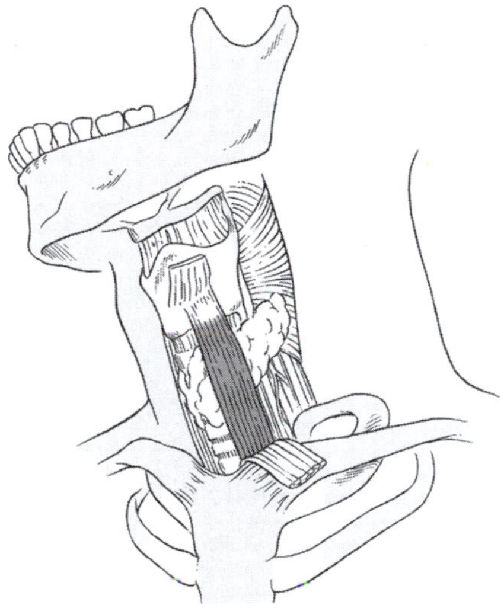

FIGURE 3.12

Muscle name ______

Origin ______

Insertion ______

Action ______

Innervation ______

FIGURE 3.9

Muscle name	Digastricus—Digastric
Origin (superior attachment)	**Posterior belly**—between the mastoid and styloid processes of temporal bone **Anterior belly**—inner side of inferior margin of mandible near mandibular symphysis
Insertion (inferior attachment)	Both bellies insert on the body of the greater cornu of the hyoid bone by a fibrous loop
Action	Acting together, the digastric muscles elevate the hyoid bone and steady it during swallowing and speech; the posterior belly helps open the mouth and depresses the mandible
Innervation	**Anterior belly**—mandibular branch of trigeminal nerve (V) **Posterior belly**—cervical branch of facial nerve (VII)

FIGURE 3.10

Muscle name	Mylohyoid(eus)
Origin (superior attachment)	Mylohyoid line of mandible
Insertion (inferior attachment)	Upper border and median raphe of hyoid bone
Action	Elevates hyoid bone and raises floor of mouth and tongue
Innervation	Mandibular branch of trigeminal nerve (V)

FIGURE 3.11

Muscle name	Sternohyoid(eus)
Origin (inferior attachment)	Medial end of clavicle and manubrium of sternum
Insertion (superior attachment)	Lower margin of body of hyoid bone
Action	Depresses hyoid bone if it has been elevated, as in swallowing
Innervation	Cervical spinal nerves C1–C3 through the ansa cervicalis (slender nerve root in cervical plexus)

FIGURE 3.12

Muscle name	Sternothyroid(eus)
Origin (inferior attachment)	Posterior surface of manubrium of sternum
Insertion (superior attachment)	Oblique line on lamina of thyroid cartilage
Action	Depresses larynx
Innervation	Ansa cervicalis (C1–C3)

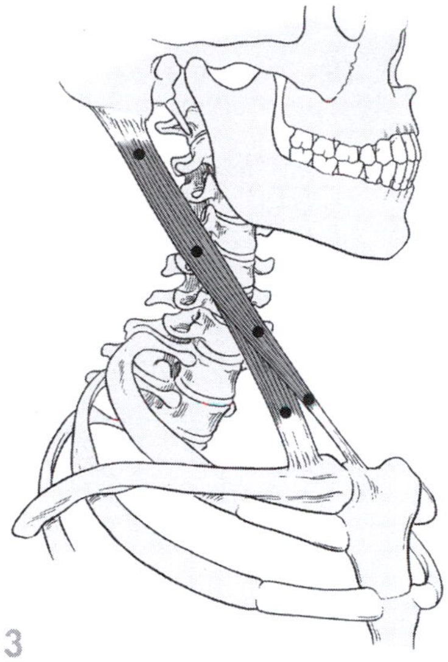

FIGURE 3.13

Muscle name ______________________

Origin ______________________

Insertion ______________________

Action ______________________

Innervation ______________________

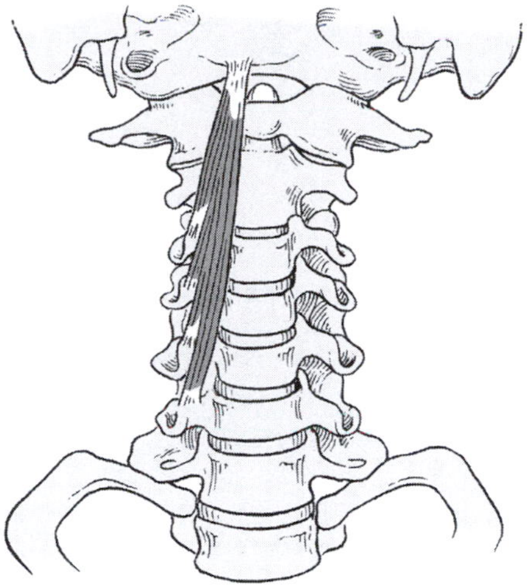

FIGURE 3.14

Muscle name ______________________

Origin ______________________

Insertion ______________________

Action ______________________

Innervation ______________________

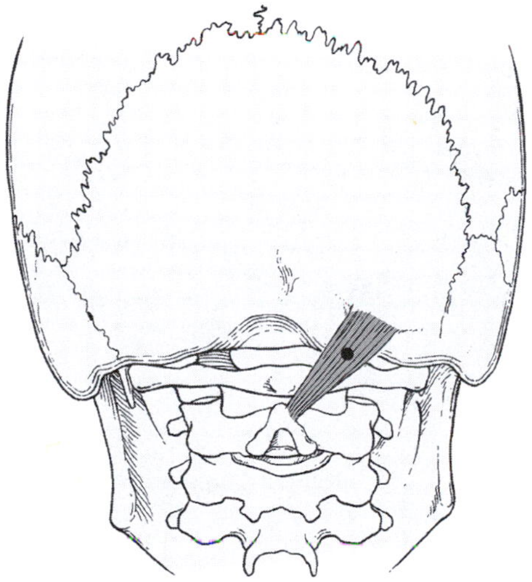

FIGURE 3.15

Muscle name ______________________

Origin ______________________

Insertion ______________________

Action ______________________

Innervation ______________________

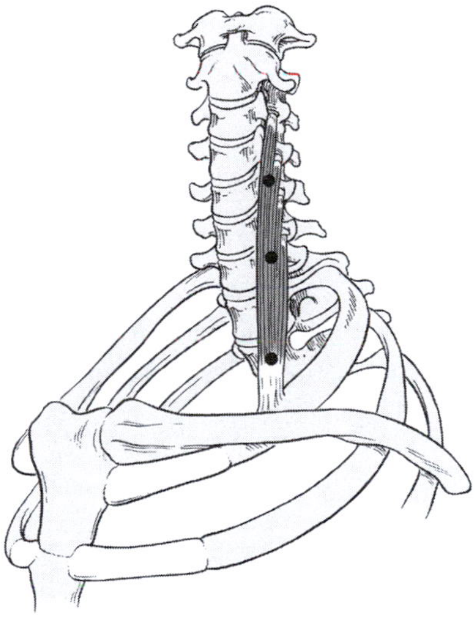

FIGURE 3.16

Muscle name ______________________

Origin ______________________

Insertion ______________________

Action ______________________

Innervation ______________________

FIGURE 3.14

Muscle name	Longus Capitis
Origin (inferior attachment)	Anterior tubercles of the transverse processes of the third through sixth cervical vertebrae
Insertion (superior attachment)	Basilar process of occipital bone anterior to foramen magnum
Action	Flexes cervical vertebrae and head
Innervation	C1–C4

FIGURE 3.16

Muscle name	Scalenus Anterior— Scalene Anterior
Origin (superior attachment)	Anterior tubercle of the transverse processes of the third through sixth cervical vertebrae
Insertion (inferior attachment)	Scalene tubercle on the inner border and upper surface of the first rib
Action	Bends the cervical portion of the vertebral column forward and laterally; it also assists in the elevation of the first rib
Innervation	Ventral rami of the fourth through sixth cervical nerves (C4–C6)

FIGURE 3.13

Muscle name	Sternocleidomastoid(eus)
Origin (inferior attachment)	Sternal head—manubrium of sternum Clavicular head—superior border of medial third of clavicle
Insertion (superior attachment)	Mastoid process of temporal and lateral half of superior nuchal line
Action	Contraction of one side—bends neck laterally and rotates head to opposite side Contraction of both sides together—flexes neck; with head fixed it assists in elevating the thorax during forced inspiration
Innervation	Spinal part of spinoaccessory nerve (XI) and branches of cervical spinal nerves (C2–C4)

FIGURE 3.15

Muscle name	Rectus Capitis Posterior Major
Origin (inferior attachment)	Spinous process of axis
Insertion (superior attachment)	Lateral portion of inferior nuchal line of occipital bone
Action	Extends and rotates the head toward the same side
Innervation	Dorsal ramus of the suboccipital nerve (C1)

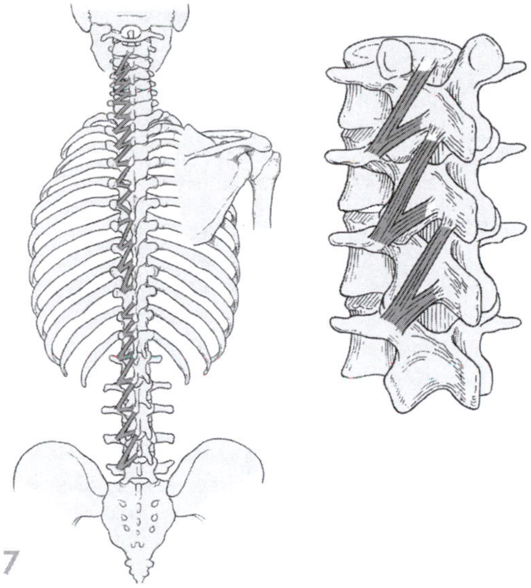

FIGURE 3.17

Muscle name ______________________

Origin ______________________

Insertion ______________________

Action ______________________

Innervation ______________________

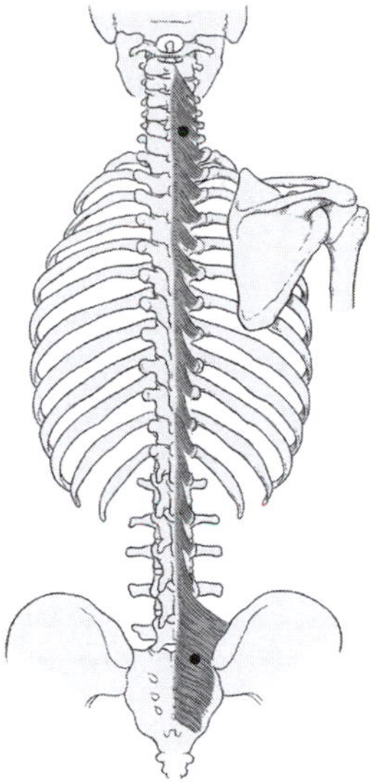

FIGURE 3.18

Muscle name ______________________

Origin ______________________

Insertion ______________________

Action ______________________

Innervation ______________________

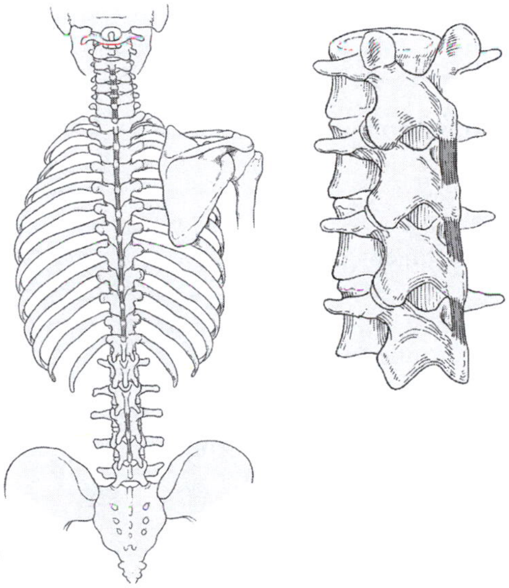

FIGURE 3.19

Muscle name ______________________

Origin ______________________

Insertion ______________________

Action ______________________

Innervation ______________________

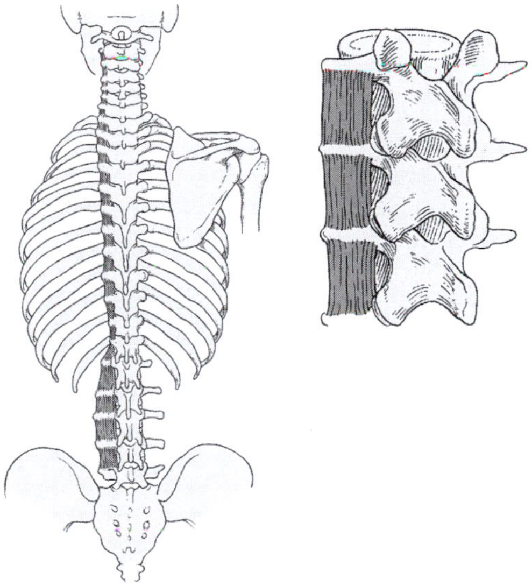

FIGURE 3.20

Muscle name ______________________

Origin ______________________

Insertion ______________________

Action ______________________

Innervation ______________________

FIGURE 3.18

Muscle name	Multifidus
Origin (inferior attachment)	Articular processes of the last four cervical, transverse processes of all thoracic, and mammillary processes of lumbar vertebrae, the posterior superior iliac spine, posterior sacroiliac ligaments, and dorsal surface of sacrum adjacent to sacral spinous processes
Insertion (superior attachment)	Spinous process of the vertebra above the vertebra of origin
Action	Extend and rotate vertebral column
Innervation	Dorsal rami of spinal nerves

FIGURE 3.17

Muscle name	Rotatores
Origin (inferior attachment)	Transverse processes of each vertebrae
Insertion (superior attachment)	**Short head**—base of spinous process of next vertebrae above **Long head**—base of spinous process of second vertebra above
Action	Extend and rotate the vertebral column
Innervation	Dorsal rami of spinal nerves

FIGURE 3.20

Muscle name	Intertransversarii
Origin (inferior attachment)	Transverse processes of all vertebrae from lumbar to axis
Insertion (superior attachment)	Transverse process of next superior vertebrae
Action	Lateral flexion of vertebral column
Innervation	Ventral and dorsal rami of spinal nerves

FIGURE 3.19

Muscle name	Interspinales
Origin (inferior attachment)	**Cervical region**—spinous processes of third to seventh cervical vertebrae (C3–C7) **Thoracic region**—spinous processes of second to twelfth thoracic vertebrae (T2–T12) **Lumbar region**—spinous processes of second to fifth lumbar vertebrae (L2–L5)
Insertion (superior attachment)	Spinous process of next superior vertebra to the vertebra of origin
Action	Extend the vertebral column
Innervation	Posterior primary rami of spinal nerves

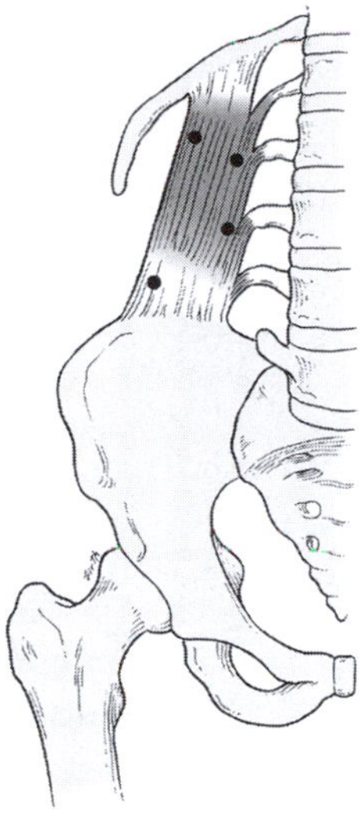

FIGURE 3.21

Muscle name ______________________

Origin ______________________

Insertion ______________________

Action ______________________

Innervation ______________________

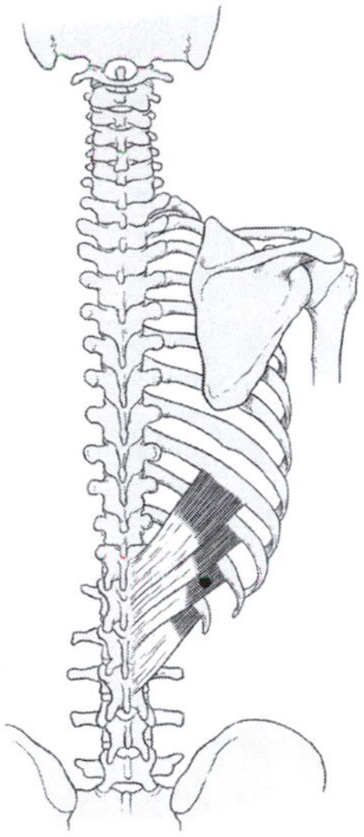

FIGURE 3.22

Muscle name ______________________

Origin ______________________

Insertion ______________________

Action ______________________

Innervation ______________________

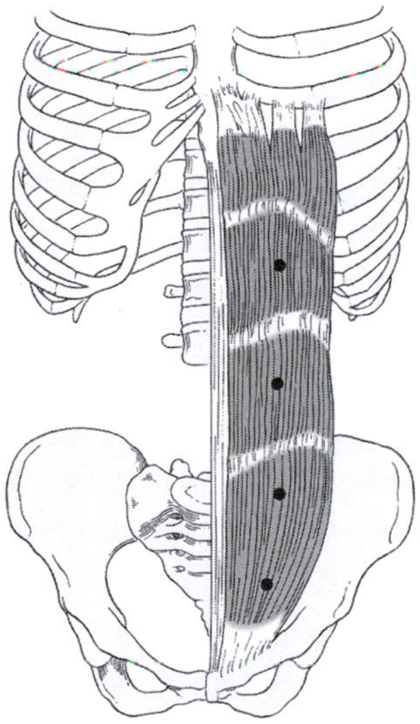

FIGURE 3.23

Muscle name ______________________

Origin ______________________

Insertion ______________________

Action ______________________

Innervation ______________________

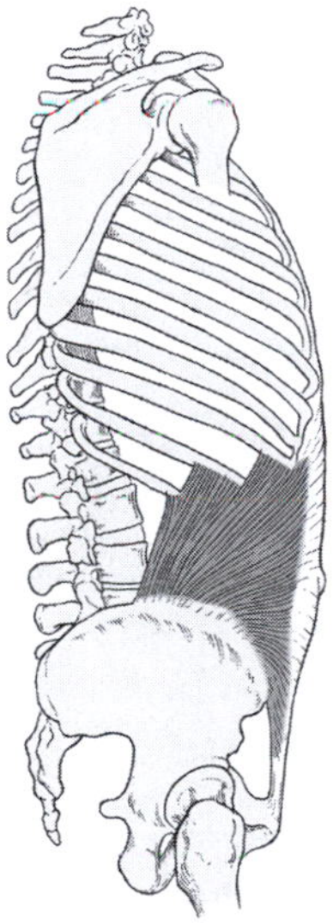

FIGURE 3.24

Muscle name ______________________

Origin ______________________

Insertion ______________________

Action ______________________

Innervation ______________________

FIGURE 3.22

Muscle name	Serratus Posterior Inferior
Origin	Spinous process of the last two thoracic and upper three lumbar vertebrae (T11–L3)
Insertion	Inferior borders and outer surfaces of lower four ribs just lateral to the angles
Action	Depresses last four ribs (this is somewhat controversial in light of recent studies since it shows no electromyographic activity during respiration)
Innervation	The ninth through twelfth thoracic nerves (T9–T12)

FIGURE 3.24

Muscle name	Obliquus Internus Abdominis—Internal Oblique
Origin	Lateral half of inguinal ligament, anterior two thirds of the iliac crest, and thoracolumbar fascia
Insertion	Upper fibers into cartilages of last three ribs, the remainder into the aponeurosis extending from the tenth costal cartilage to the pubic bone
Action	Compresses abdominal contents, laterally bends and rotates vertebral column; it also aids the rectus abdominis in flexing vertebral column
Innervation	Ventral rami of the lower six thoracic and first lumbar spinal nerves (T7–T12, L1)

FIGURE 3.21

Muscle name	Quadratus Lumborum
Origin (inferior attachment)	Iliolumbar ligament and the posterior portion of the iliac crest
Insertion (superior attachment)	Inferior border of last rib and the transverse processes of the first four lumbar vertebrae (L1–L4)
Action	Flexes lumbar region of vertebral column laterally to the same side. Both muscles together stabilize and extend the lumbar vertebrae and assist forced expiration.
Innervation	Ventral rami of the twelfth thoracic (T12) and upper three lumbar spinal nerves (L1–L3)

FIGURE 3.23

Muscle name	Rectus Abdominis
Origin (inferior attachment)	Crest of pubis and pubic symphysis
Insertion (superior attachment)	Cartilage of fifth, sixth, and seventh ribs and xiphoid process of sternum
Action	Compresses the abdominal cavity and flexes the vertebral column
Innervation	Anterior primary rami of the seventh through twelfth intercostal nerves (T7–T12)

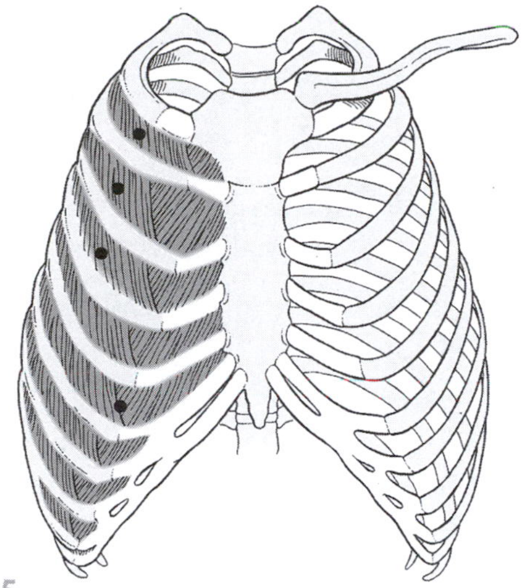

FIGURE 3.25

Muscle name ______________________

Origin ______________________

Insertion ______________________

Action ______________________

Innervation ______________________

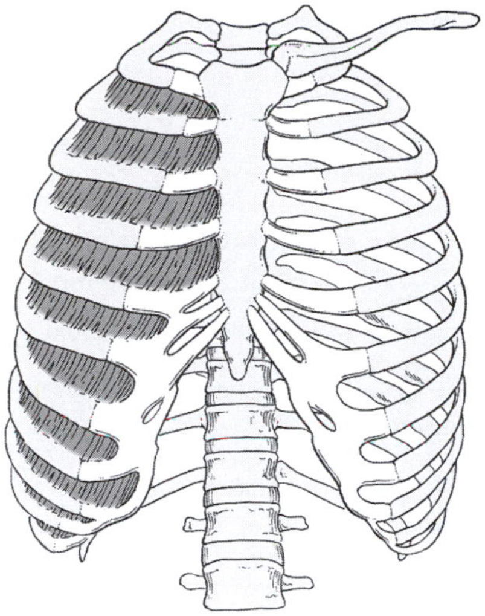

FIGURE 3.26

Muscle name ______________________

Origin ______________________

Insertion ______________________

Action ______________________

Innervation ______________________

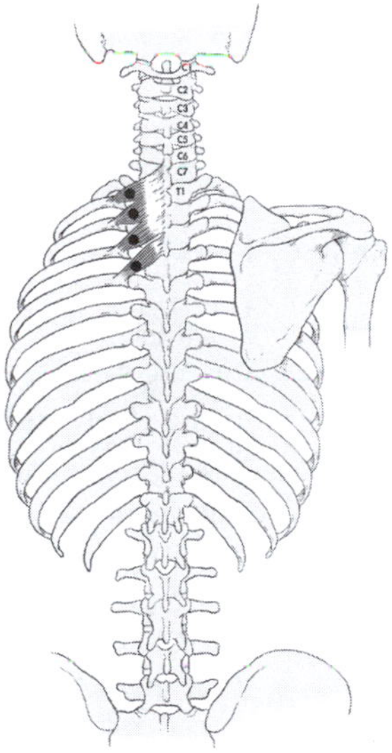

FIGURE 3.27

Muscle name ______________________

Origin ______________________

Insertion ______________________

Action ______________________

Innervation ______________________

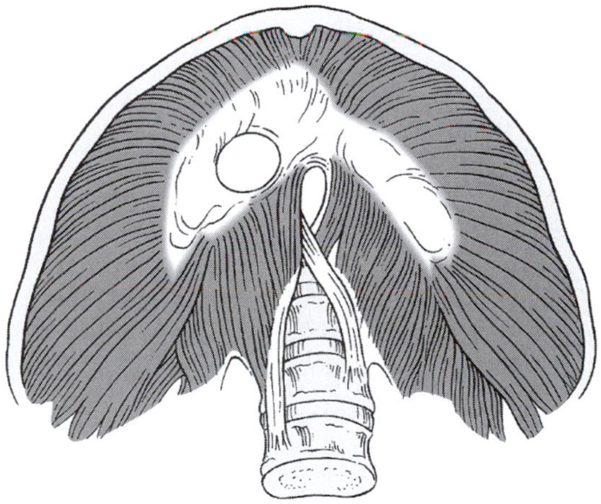

FIGURE 3.28

Muscle name ______________________

Origin ______________________

Insertion ______________________

Action ______________________

Innervation ______________________

FIGURE 3.26

Muscle name	Intercostales Interni—Internal Intercostals
Origin (superior attachment)	Ridge of inner surface of rib and corresponding costal cartilage
Insertion (inferior attachment)	Superior border of rib below
Action	Draw ribs together and depress the rib cage
Innervation	Intercostal nerves (T1–T11)

FIGURE 3.25

Muscle name	Intercostales Externi—External Intercostals
Origin (superior attachment)	Lower margin of upper eleven ribs
Insertion (inferior attachment)	Superior border of rib below
Action	With first ribs fixed by scalenes, they pull the ribs toward one another to elevate the rib cage.
Innervation	Intercostal nerves (T1–T11)

FIGURE 3.28

Muscle name	Diaphragm
Origin	First three lumbar vertebrae, lower six costal cartilages, and inner surface of xiphoid process of sternum
Insertion	Muscle fibers converge upward and inward to form the central tendon
Action	Flattens on contraction, increasing the vertical dimensions of thorax
Innervation	Phrenic nerve (C3–C5)

FIGURE 3.27

Muscle name	Serratus Posterior Superior
Origin	Lower portion of ligamentum nuchae and the spinous processes of the sixth and seventh cervical through the third thoracic vertebrae (C6–T3)
Insertion	Upper border and external surfaces of ribs two through five lateral to their angles
Action	Assists in raising ribs during inspiration
Innervation	Second through fifth intercostal nerves (T2–T5)

II. MATCHING

Group A: In the blank to the left of each muscle action, write the **LETTER** of the head muscle that causes that action.

1. _____ Large muscle that elevates and protracts mandible
2. _____ Puckers lips
3. _____ Draws eyebrows together as in frowning
4. _____ Draws angle of lip upward and outward as in smiling
5. _____ Draws back scalp and wrinkles forehead
6. _____ Draws corner of mouth laterally and compresses cheek as in sucking through a straw
7. _____ Closes eyelids
8. _____ Elevates upper lip and makes nasolabial furrow
9. _____ Elevates and retracts mandible, assists side-to-side movement of mandible
10. _____ Raises upper eyelid

A. Occipitofrontalis
B. Orbicularis oculi
C. Levator palpebrae superioris
D. Corrugator supercilii
E. Orbicularis oris
F. Levator labii superioris
G. Zygomaticus major
H. Buccinator
I. Temporalis
J. Masseter

Group B: In the blank to the left of each origin description, write the **LETTER** of the neck muscle with that origin.

1. _____ Manubrium of sternum and medial one-third of clavicle
2. _____ Thyroid cartilage at oblique line
3. _____ Anterior base of transverse process of atlas
4. _____ **Posterior belly:** between mastoid and styloid processes
   Anterior belly: inferior margin of mandible near mandibular symphysis
5. _____ Anterior tubercle of transverse processes of cervical vertebrae 3–6
6. _____ Mylohyoid line of mandible
7. _____ Inferior mental spine of mandible
8. _____ Styloid process of temporal bone

A. Digastric
B. Mylohyoid
C. Stylohyoid
D. Geniohyoid
E. Thyrohyoid
F. Sternocleidomastoid
G. Longus capitis
H. Rectus capitis anterior

Group C: In the blank to the left of each muscle action, write the **LETTER** of the trunk muscle that causes that action.

1. ______ Flexes lumbar region of vertebral column laterally
2. ______ Compresses abdominal cavity and flexes vertebral column
3. ______ Group of muscles that extend and laterally flex vertebral column
4. ______ Compresses abdominal cavity and laterally flexes and rotates vertebral column
5. ______ Elevates rib cage; used in inspiration
6. ______ Extends and hyperextends the head; contraction of one side laterally flexes neck
7. ______ Draws ribs together and depresses rib cage
8. ______ Draws ventral part of rib down
9. ______ Deep muscles that extend and rotate vertebral column
10. ______ Raises ribs and laterally flexes and rotates vertebral column

A. Splenius capitis
B. Multifidus
C. Erector spinae
D. Quadratus lumborum
E. Levatores costarum
F. External intercostals
G. Transversus thoracis
H. Internal intercostals
I. Rectus abdominis
J. External obliques

III. SENTENCE COMPLETION

Circle the term in parentheses that correctly completes each statement.

1. The muscle that is absent in some people that functions to raise the ears is the (**temporalis, temporoparietalis, occipitofrontalis, procerus**).
2. The muscle that closes the eyelids is the (**orbicularis oris, levator palpebrae superioris, orbicularis oculi, procerus**).
3. The facial muscle used in frowning is the (**procerus, orbicularis oculi, corrugator supercilii, depressor septi**).
4. The muscles of facial expression such as those in the previous questions are innervated by cranial nerve (**III, IV, V, VII**).
5. All of the following muscles are involved in smiling *except* the (**levator labii superioris, risorius, levator anguli oris, zygomaticus major**).
6. The muscle involved in both sucking through a straw and blowing air out as in playing a trumpet is the (**buccinator, mentalis, risorius, temporalis**).
7. The jaw muscle that bulges when you clench your teeth is the (**temporalis, masseter, medial pterygoid, buccinator**).
8. The broad flat muscle that covers the neck and tenses the skin on the neck is the (**digastric, procerus, platysma, sternocleidomastoid**).
9. This neck muscle's name means "two bellied." It is the (**digastric, biceps, epigastric, diceps**).
10. Several muscles aid in swallowing. All of the following muscles aid in swallowing by raising the floor of the mouth and the tongue *except* the (**sternohyoid, stylohyoid, mylohyoid, geniohyoid**).

11. All of the following muscles depress the hyoid *except* the (**sternohyoid, stylohyoid, thyrohyoid, omohyoid**).

12. Several muscles work together to flex the neck anteriorly. All of the following muscles *except* the (**sternocleidomastoid, longus capitis, rectus capitis anterior, obliquus capitis superior**) flex the neck anteriorly.

13. Several muscles work together to extend the neck. All of the following muscles *except* the (**scalenes medius, splenius capitis, semispinalis capitis, splenius cervis**) extend the neck.

14. The erector spinae group is made up of several muscles. All of the following *except* the (**multifidus, iliocostalis, longissimus, spinalis**) are part of the erector spinae group.

15. Several muscles together extend and rotate the vertebral column. All of the following *except* the (**multifidus, semispinalis, rotares, intertransversarii**) extend and rotate the vertebral column.

16. A major muscle that flexes the lumbar region of the vertebral column laterally is the (**levatores costarum, quadratus lumborum, iliacus, psoas major**).

17. All of the following muscles *except* the (**external oblique, rectus abdominis, rectus femoris, internal oblique**) compress the abdominal cavity and flex the vertebral column.

18. Synergists of the serratus posterior superior include all of the following *except* the (**serratus posterior inferior, scalenes group, external intercostals, levatores costarum**).

19. The diaphragm is innervated by the (**vagus, phrenic, spinoaccessory, intercostal**) nerve.

20. The muscles that raise the ribs during inspiration are the (**external intercostals, internal intercostals, subcostales, transverses thoracis**).

IV. MULTIPLE CHOICE

In the blank to the left of each question, write the **LETTER** of the correct answer.

1. _____ The splenius muscles extend and hyperextend the neck. Which of the following muscles is antagonistic to the splenius muscles?
   A. Sternocleidomastoid
   B. Semispinalis
   C. Iliocostalis
   D. Levatores costarum

2. _____ Which major muscle group is antagonistic to the rectus abdominis muscle?
   A. Splenius
   B. Levatores costarum
   C. Erector spinae
   D. Intertransversarii

3. _____ Which muscle group is responsible for lateral flexion of the vertebral column?
   A. Interspinales
   B. Multifidus
   C. Spinalis
   D. Intertransversarii

4. _____ Which of the following muscles are antagonistic to the external intercostals?
   A. Levatores costarum
   B. Splenius
   C. Internal intercostals
   D. Subclavius

5. _____ A large muscle extends between the iliac crest and the transverse processes of the lumbar vertebrae and flexes the lumbar region laterally. Name this muscle.
   A. External oblique
   B. Quadratus lumborum
   C. Cremaster
   D. Transversus abdominis

6. _____ Several muscles work synergistically to pull down the ribs. Which of the following muscles does *not* pull down the ribs?
   A. Transversus thoracis
   B. Subcostales
   C. Levatores costarum
   D. Internal intercostals

7. _____ Which of the following muscles does *not* assist in forced expiration?
   A. Internal oblique
   B. Internal intercostals
   C. Serratus posterior superior
   D. Diaphragm

8. _____ Which cranial nerve innervates the muscles of facial expression?
   A. Oculomotor
   B. Trigeminal
   C. Abducens
   D. Facial

9. _____ Which of the four rectus muscles of the eye is *not* innervated by the oculomotor nerve?
   A. Superior rectus
   B. Inferior rectus
   C. Lateral rectus
   D. Medial rectus

10. _____ Which of the following muscles encircles the mouth, closing it and causing the lips to protrude?
   A. Orbicularis oculi
   B. Orbicularis oris
   C. Levator palpebrae
   D. Mentalis

11. _____ The zygomaticus major elevates the upper lip upward and outward. Which of the following muscles is antagonistic to the zygomaticus major in elevation of upper lip?
   A. Risorius
   B. Levator anguli oris
   C. Depressor anguli oris
   D. Zygomaticus minor

12. _____ Which muscle is one of two that are largely responsible for the side-to-side motions involved in chewing?
   A. Medial pterygoid
   B. Masseter
   C. Buccinator
   D. Temporoparietalis

13. _____ Which of the following muscles are you using if you can raise your ears?
   A. Occipitofrontalis
   B. Temporalis
   C. Procerus
   D. Temporoparietalis

14. _____ Which of the following muscles is antagonistic to the digastric?
   A. Temporalis
   B. Mylohyoid
   C. Geniohyoid
   D. Platysma

15. _____ Which of these muscles is considered an infrahyoid muscle?
   A. Mylohyoid
   B. Stylohyoid
   C. Geniohyoid
   D. Sternohyoid

16. _____ Which of these anterior neck muscles is most superficial?
   A. Geniohyoid
   B. Thyrohyoid
   C. Sternohyoid
   D. Sternothyroid

17. _____ Which of these small posterior neck muscles is *not* antagonistic to the rectus capitis anterior?
   A. Rectus capitis posterior major
   B. Obliquus capitis superior
   C. Longus colli
   D. Obliquus capitis inferior

18. _____ Which of the following muscles laterally flexes the neck and helps to raise the first rib?

A. Scalenus medius
B. Rectus capitis lateralis
C. Longus capitis
D. Longus colli

19. _____ Which of the following terms means "straight"?

A. Pinnate
B. Rectus
C. Palmate
D. Convergent

20. _____ Which is the deepest abdominal muscle?

A. External oblique
B. Internal oblique
C. Transverse abdominis
D. Rectus abdominis

V. FILL IN THE BLANK

Write the correct word(s) in the blank spaces to complete each statement.

1. The Galea aponeurotica connects the ____________________ and ____________________ muscles.
2. The muscle that raises the upper eyelid is the ____________________ .
3. The muscle that wrinkles the skin on the bridge of the nose is the ____________________ .
4. The muscle that maintains the opening of the nostrils of the nose during forceful inspiration is the ____________________ .
5. The two major muscles that raise the jaw are the ____________________ and ____________________ .
6. The two deep muscles that produce the side-to-side grinding motion are the ____________________ and ____________________ .
7. The name of all the extrinsic muscles of the tongue include the root word ____________________ .
8. The two attachments of the stylohyoideus muscle are the ____________________ and the ____________________. A muscle synergistic with it is the ____________________ .
9. Two muscles that depress the hyoid during swallowing are the ________________ and ________________ .
10. The major muscle that flexes the neck and, when only one side is contracted, rotates the head is the ____________________. Its two attachments are the ________________ and ________________.

11. A short neck muscle that has attachments on the transverse process of the atlas and the occipital bone and that aids in flexing the head is the ______________________ .

12. Two short neck muscles that have attachments on the atlas and axis and the occipital bone extend the head. They are the ______________________ and ______________________ .

13. The erector spinae muscles include three groups of muscles located along the vertebral column. They are the ______________________ , ______________________ , and ______________________ groups.

14. A muscle made of several slips that raises the ribs and laterally flexes and rotates the vertebral column is the ______________________ .

15. The thoracic muscles that depress the rib cage and are important in forced exhalation include the ______________________ , ______________________ , and ______________________ .

VI. SHORT ANSWER

Answer each of the questions in the space provided.

1. What symptoms would lead to a diagnosis of Bell's palsy?

____

____

____

2. Why is it necessary for the anesthesiologist to insert an air tube during surgery?

____

____

____

3. A patient cannot flex his neck. Which muscles might be paralyzed?

____

____

____

4. Explain what is meant by back strain. Which muscles are most likely involved? What are some common causes of back strain?

5. What muscles are antagonistic to the rectus abdominis muscle?

VII. CASE STUDIES

For each case study, read the paragraph and answer the associated questions in the space provided.

A. During a rear-end collision, a 60-year-old woman driver was thrown forward at the same time her head was thrown back. She suffered a concussion. During her examination at the doctor's office, she complained of a stiff and painful neck. The doctor noted that the woman held her head rigidly and slightly flexed and turned to the right. She had a weak biceps reflex on the right side. Results of a cervical X-ray showed that the disc between C-5 and C-6 was thin.

   1. What type of injury caused these symptoms?

   2. Which group of spinal nerves was probably compressed?

   3. Because the woman's head was flexed, which group of muscles would have been stretched and possibly torn?

4. The weak biceps reflex would have been caused by compression of which nerve?

B. A landscape gardener carrying heavy rocks while building a wall suddenly experienced severe pain in his lower back. Later he developed leg problems, including pain in the posterior lateral right thigh, calf, and foot. He limped when he walked. He still had severe pain in the lower back, and he lacked an ankle reflex. X-rays showed a protrusion of the intervertebral disc between L5 and S1.

1. Which lower back muscles were undergoing spasms?

2. Which spinal nerve was being compressed?

3. Which leg muscles are innervated by this nerve?

4. Why was the patient limping?

Muscles of the Shoulder, Arm, and Hand

4

At the end of this chapter, you should be able to

1. Identify the muscles of the shoulder, arm, and hand.
2. Name the attachments for the major shoulder, arm, and hand muscles.
3. Name the actions for the major shoulder, arm, and hand muscles.
4. Identify the four muscles of the rotator cuff.
5. Explain golfer's elbow.
6. Explain Poland's syndrome.
7. Explain carpal tunnel syndrome
8. Identify the muscles involved in supination and pronation.
9. Describe the anatomical snuff box.

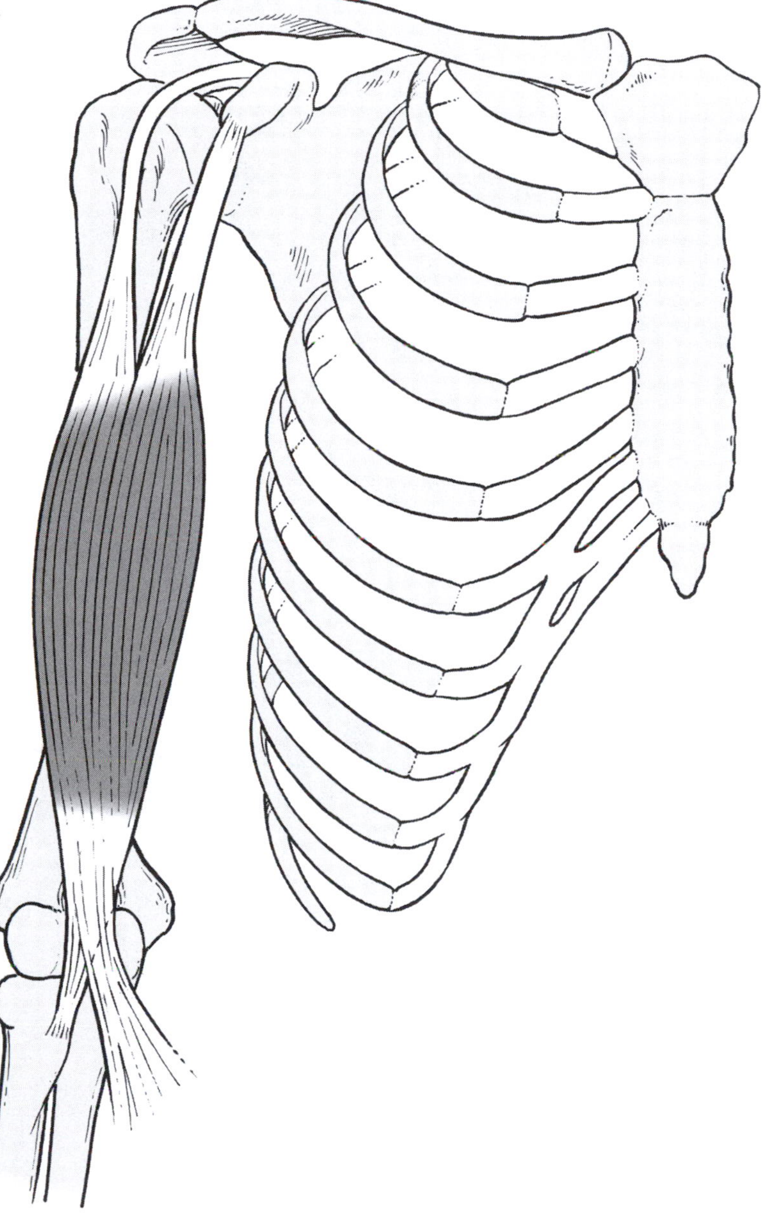

Word Search Puzzle

INSTRUCTIONS: Find and CIRCLE each of the listed terms in the Word Search Puzzle.
(Terms may read from left to right, right to left, up, down, or diagonally.)

S Z B R A C H I O R A D I A L I S H G F D
U X I Q B V W Q X W X Q T N T R I C E P S
I V C Z D E L T O I D C R C R V R G F P U
V B E W U W X R P B Q S P O X Q A H Q X P
A Z P V C F P O P C C F X N L B M Y W V R
L K S Y T Z I T O D F E R E P F L Z B C A
C F B X O P X A N X L D O U L N A R I S S
B B R D R Q W N E F X M T S A Q P G F D P
U X A A Z Q V I N S T U A W T E R E S H I
S I C I L L O P S B D R N D I G I T I P N
C Z H F W X U U K G H O O F S I V E R B A
X B I S X R T S F S R T R P S Z W V S Q T
Z S I L A R O T C E P I P F I V X Y Z B U
Q Z P C F L F B D G H G H G M W H F G D S
Q X I E S S O R E T N I B C U X J X D B A
V W Y Y R H O M B O I D E U S Y X Z J K L

ABDUCTOR
ANCONEUS
BICEPS
BRACHII
BRACHIORADIALIS
BREVIS
CARPI
DELTOID

DIGITI
DIGITORUM
FLEXOR
INTEROSSEI
LATISSIMUS
OPPONENS
PALMARIS
PECTORALIS

POLLICIS
PRONATOR
RHOMBOIDEUS
SUBCLAVIUS
SUPINATOR
SUPRASPINATUS
TERES
TRICEPS
ULNARIS

I. ILLUSTRATION IDENTIFICATION (pages 71–84 can be cut and used as flash cards)

Write the technical information for each muscle in the spaces beneath the illustrations.

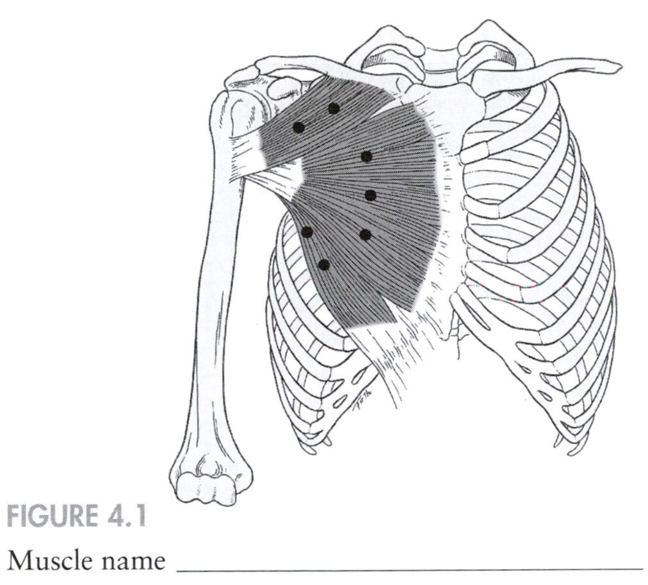

FIGURE 4.1

Muscle name ______________________________

Origin ______________________________

Insertion ______________________________

Action ______________________________

Innervation ______________________________

FIGURE 4.2

Muscle name ______________________________

Origin ______________________________

Insertion ______________________________

Action ______________________________

Innervation ______________________________

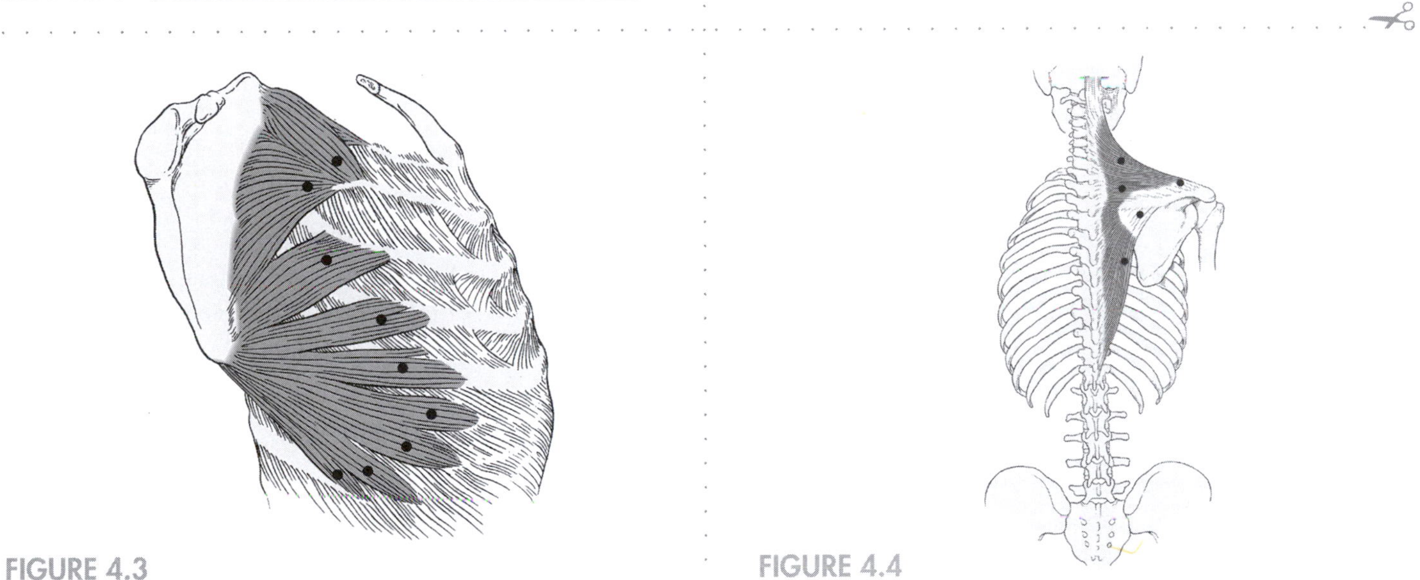

FIGURE 4.3

Muscle name ______________________________

Origin ______________________________

Insertion ______________________________

Action ______________________________

Innervation ______________________________

FIGURE 4.4

Muscle name ______________________________

Origin ______________________________

Insertion ______________________________

Action ______________________________

Innervation ______________________________

FIGURE 4.2

Muscle name	Pectoralis Minor
Origin (proximal attachment)	Anterior surfaces of the third through fifth rib
Insertion (distal attachment)	Coracoid process of the scapula
Action	With ribs fixed, it draws the scapula forward and downward, and with scapula fixed, it draws the rib cage superiorly
Innervation	Medial pectoral nerve (C8, T1)

FIGURE 4.1

Muscle name	Pectoralis Major
Origin (proximal attachment)	Ventral surface of the sternum down to the seventh rib, sternal half of clavicle, cartilage of true ribs, and aponeurosis of the external oblique muscle
Insertion (distal attachment)	Lateral lip of the intertubercular groove of the humerus
Action	Protracts scapula, adducts and medially rotates arm, clavicular head flexes humerus, sternal head extends humerus, and with insertion fixed it assists in elevation of the thorax
Innervation	Medial and lateral pectoral nerves (C5–C8, T1)

FIGURE 4.4

Muscle name	Trapezius
Origin (proximal attachment)	Medial third of superior nuchal line, external occipital protuberance, ligamentum nuchae, spinous process of the seventh cervical, and all thoracic vertebrae
Insertion (distal attachment)	Lateral third of clavicle, acromion process, and spine of scapula
Action	Stabilizes, raises, retracts, and rotates the scapula. The superior fibers elevate, the middle fibers retract, and the inferior fibers depress the scapula. Together the upper and lower fibers rotate the arm.
Innervation	Spinoaccessory (XI) and ventral rami of C3, C4

FIGURE 4.3

Muscle name	Serratus Anterior
Origin (proximal attachment)	Outer surfaces and superior borders of the first eight or nine ribs
Insertion (distal attachment)	Anterior surface of the medial border of the scapula
Action	Abducts, protracts, and upwardly rotates the scapula; it also holds the scapula firmly against the thorax
Innervation	Long thoracic nerve (C5–C7)

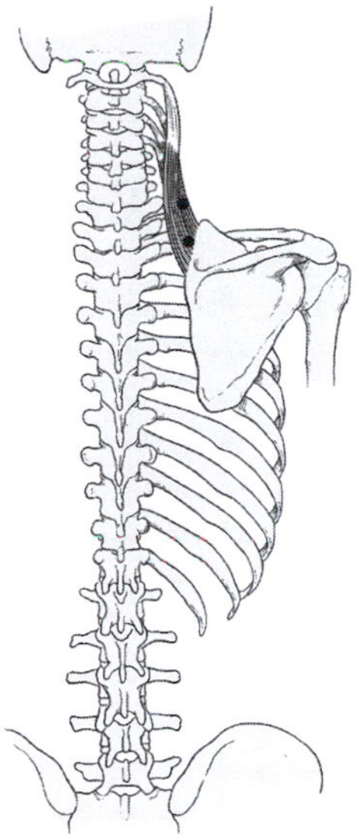

FIGURE 4.5

Muscle name ______________________

Origin ______________________

Insertion ______________________

Action ______________________

Innervation ______________________

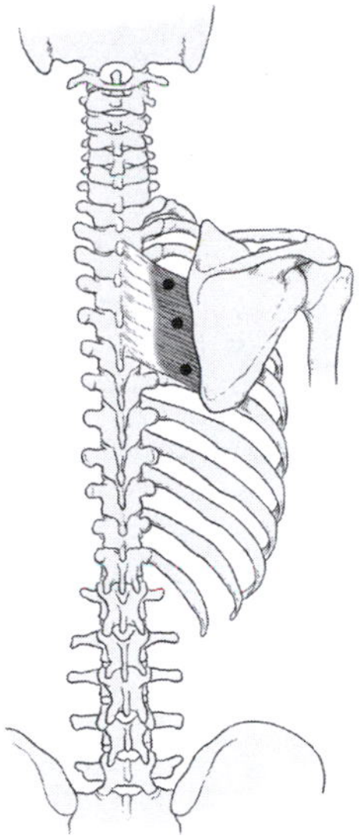

FIGURE 4.6

Muscle name ______________________

Origin ______________________

Insertion ______________________

Action ______________________

Innervation ______________________

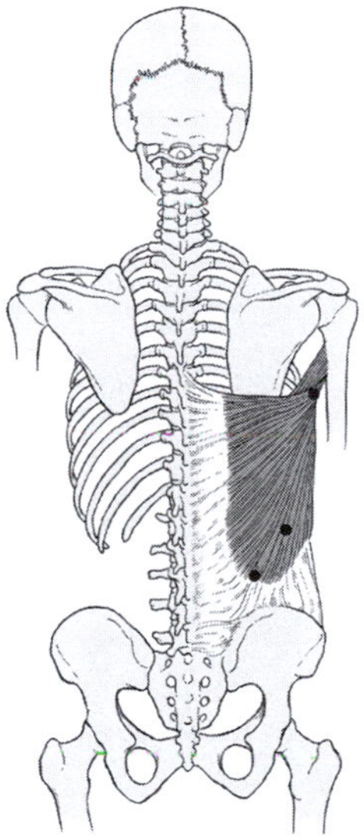

FIGURE 4.7

Muscle name ______________________

Origin ______________________

Insertion ______________________

Action ______________________

Innervation ______________________

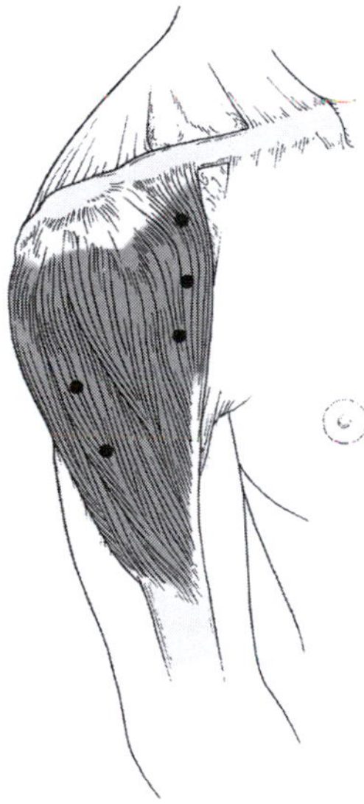

FIGURE 4.8

Muscle name ______________________

Origin ______________________

Insertion ______________________

Action ______________________

Innervation ______________________

FIGURE 4.6

Muscle name	Rhomboid(eus) Major
Origin (medial attachment)	Spinous process of the second through fifth thoracic vertebrae (T2–T5)
Insertion (lateral attachment)	Medial border of scapula between the spine and the inferior angle
Action	Adducts, retracts, elevates, and rotates the scapula so that the glenoid cavity faces downward and stabilizes the scapula
Innervation	Dorsal scapular nerve (C5)

FIGURE 4.8

Muscle name	Deltoid(eus)
Origin (proximal/ medial)	**Anterior portion**—superior surface of lateral third of clavicle **Middle portion**—lateral border of acromion process of scapula **Posterior portion**—lower border of the crest of the spine of the scapula
Insertion (distal/lateral)	Deltoid tuberosity of the humerus
Action	**Anterior portion**—flexion and medial rotation of the arm **Middle portion**—abducts the arm **Posterior portion**—extends and laterally rotates the arm
Innervation	Axillary nerve (C5, C6)

FIGURE 4.5

Muscle name	Levator Scapulae
Origin (superior/ medial attachment)	Transverse processes of the first four cervical vertebrae (C1–C4)
Insertion (inferior/ lateral attachment)	Vertebral border of the scapula between the superior angle and the spine
Action	Raises scapula and draws it medially; with the scapula fixed, bends the neck laterally and rotates it to the same side
Innervation	Third and fourth cervical spinal nerves and dorsal scapular nerve (C3–C5)

FIGURE 4.7

Muscle name	Latissimus Dorsi
Origin	Indirect attachment through lumbodorsal fascia into spinous process of lower six thoracic and lumbar vertebrae (T7–L5), lower three to four ribs, and iliac crest
Insertion	In floor of intertubercular groove of humerus
Action	Extends, adducts, and medially rotates the arm; draws the shoulder downward and backward
Innervation	Thoracodorsal nerve (C6–C8)

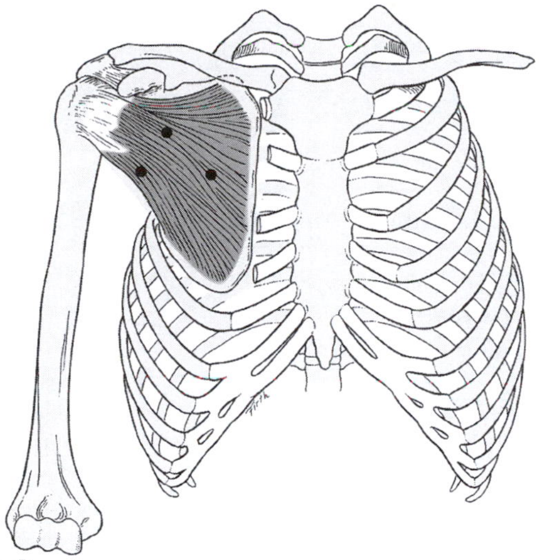

FIGURE 4.9

Muscle name ______________________

Origin ______________________

Insertion ______________________

Action ______________________

Innervation ______________________

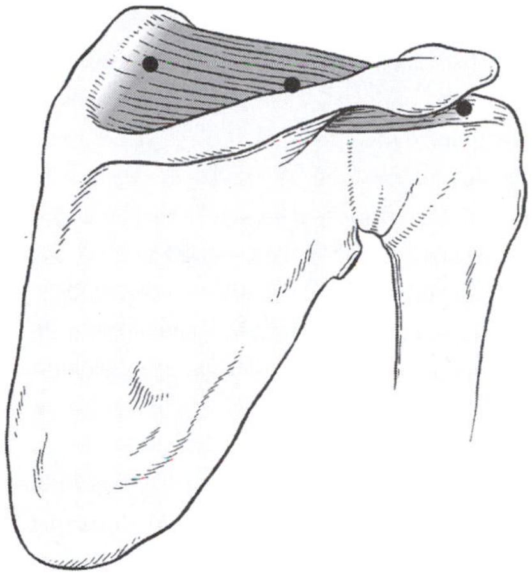

FIGURE 4.10

Muscle name ______________________

Origin ______________________

Insertion ______________________

Action ______________________

Innervation ______________________

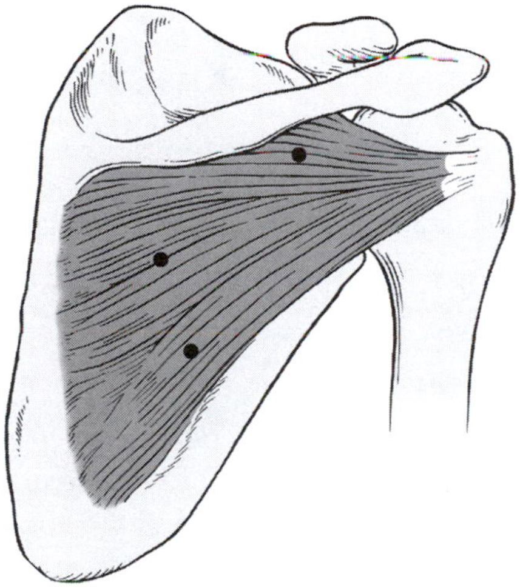

FIGURE 4.11

Muscle name ______________________

Origin ______________________

Insertion ______________________

Action ______________________

Innervation ______________________

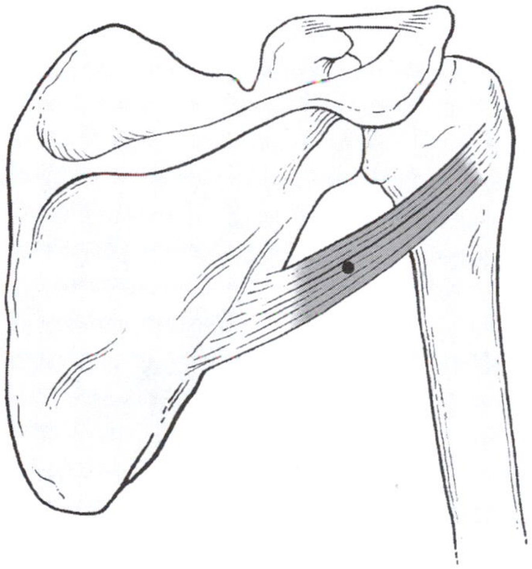

FIGURE 4.12

Muscle name ______________________

Origin ______________________

Insertion ______________________

Action ______________________

Innervation ______________________

FIGURE 4.10

Muscle name	Supraspinatus
Origin (proximal/ medial)	Supraspinous fossa of the scapula
Insertion (distal/lateral)	Superior part of the greater tubercle of the humerus and the capsule of the shoulder joint
Action	Abducts the arm and acts to stabilize the humeral head in the glenoid cavity during movements of the shoulder joint
Innervation	Suprascapular nerve (C5, C6)

FIGURE 4.9

Muscle name	Subscapularis
Origin (proximal/ medial)	Subscapular fossa of the scapula
Insertion (distal/lateral)	Lesser tubercle of the humerus and the ventral part of the capsule of the shoulder joint
Action	Medially rotates and stabilizes the head of the humerus in the glenoid cavity
Innervation	Upper and lower subscapular nerves (C5, C6)

FIGURE 4.12

Muscle name	Teres Minor
Origin (proximal/ medial)	Upper two thirds of the dorsal surface of the axillary border of the scapula
Insertion (distal/lateral)	The capsule of the shoulder joint and the inferior part of the greater tubercle of the humerus
Action	Laterally rotates the arm and draws the humerus toward the glenoid cavity
Innervation	Axillary nerve (C5)

FIGURE 4.11

Muscle name	Infraspinatus
Origin (proximal/ medial)	Infraspinous fossa of the scapula
Insertion (distal/lateral)	Middle part of the greater tubercle of the humerus and the capsule of the shoulder joint
Action	Lateral rotation of the shoulder and acts to stabilize the humeral head in the glenoid cavity; it abducts the humerus.
Innervation	Suprascapular nerve (C5, C6)

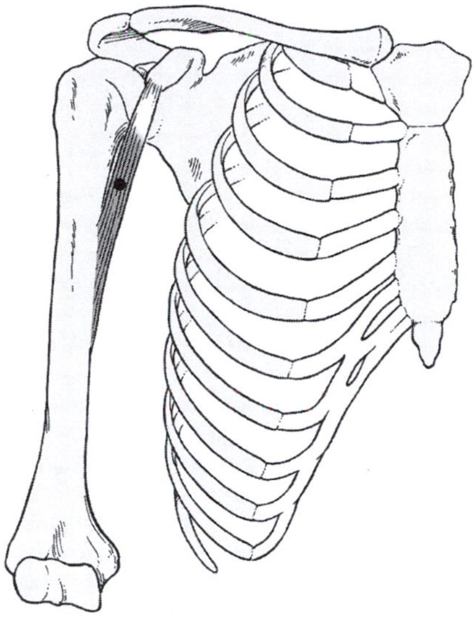

FIGURE 4.13

Muscle name ______________________

Origin ______________________

Insertion ______________________

Action ______________________

Innervation ______________________

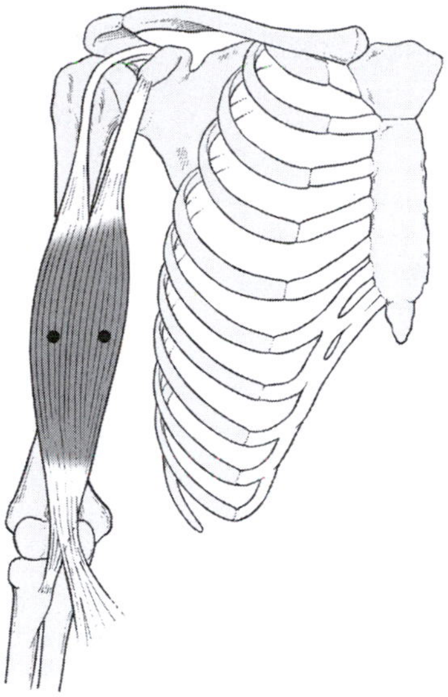

FIGURE 4.14

Muscle name ______________________

Origin ______________________

Insertion ______________________

Action ______________________

Innervation ______________________

FIGURE 4.15

Muscle name ______________________

Origin ______________________

Insertion ______________________

Action ______________________

Innervation ______________________

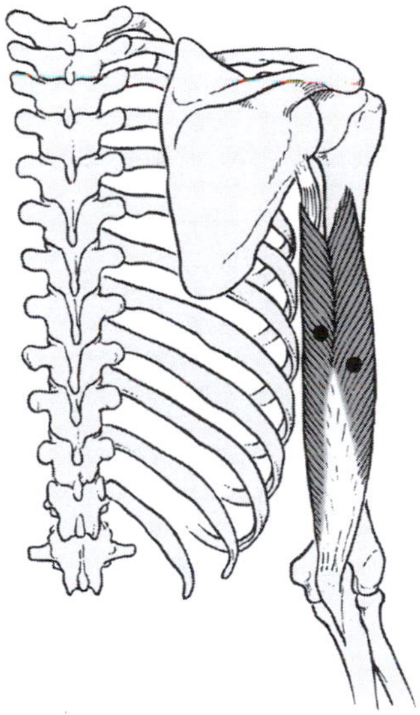

FIGURE 4.16

Muscle name ______________________

Origin ______________________

Insertion ______________________

Action ______________________

Innervation ______________________

FIGURE 4.14

Muscle name	Biceps Brachii
Origin (proximal attachment)	**Long head**—supraglenoid tubercle of the scapula **Short head**—tip of the Coracoid process of the scapula
Insertion (distal attachment)	Radial tuberosity and into aponeurosis of flexor muscles of lower arm
Action	Flexes and supinates the forearm; flexes the humerus
Innervation	Musculocutaneous nerve (C5, C6)

FIGURE 4.16

Muscle name	Triceps Brachii
Origin (proximal attachment)	**Long head**—infraglenoid tubercle of the scapula **Medial head**—distal two-thirds of the medial and posterior surfaces of the humerus **Lateral head**—upper half of the posterior surface of the humerus
Insertion (distal attachment)	Posterior surface of the olecranon process of the ulna
Action	Extends the forearm and the tendon of the long head helps stabilize the shoulder joint and extends the humerus
Innervation	Radial nerve (C7, C8)

FIGURE 4.13

Muscle name	Coracobrachialis
Origin (proximal attachment)	Tip of coracoid process of scapula
Insertion (distal attachment)	Anteromedial surface of the middle of the humerus shaft opposite the deltoid tuberosity
Action	Flexion and adduction of the humerus
Innervation	Musculocutaneous nerve (C5–C7)

FIGURE 4.15

Muscle name	Brachialis
Origin (proximal attachment)	Distal half of the anterior surface of the humerus
Insertion (distal attachment)	Coronoid process and tuberosity of ulna
Action	Flexes elbow
Innervation	Musculocutaneous and radial nerves (C5, C6)

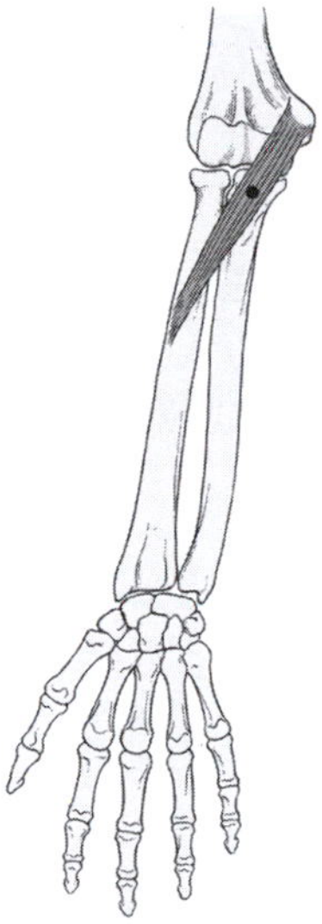

FIGURE 4.17

Muscle name ______________________

Origin ______________________

Insertion ______________________

Action ______________________

Innervation ______________________

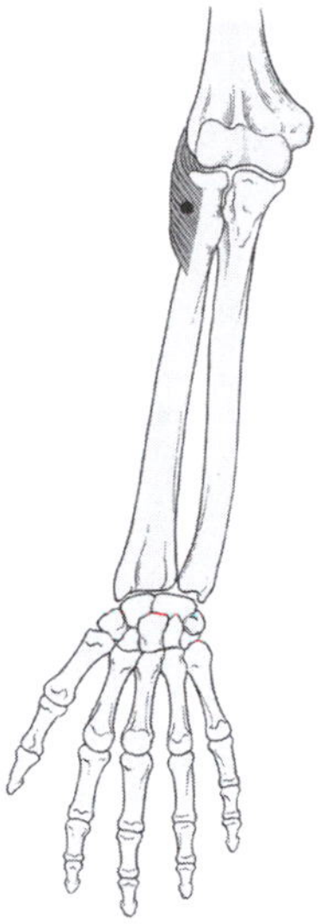

FIGURE 4.18

Muscle name ______________________

Origin ______________________

Insertion ______________________

Action ______________________

Innervation ______________________

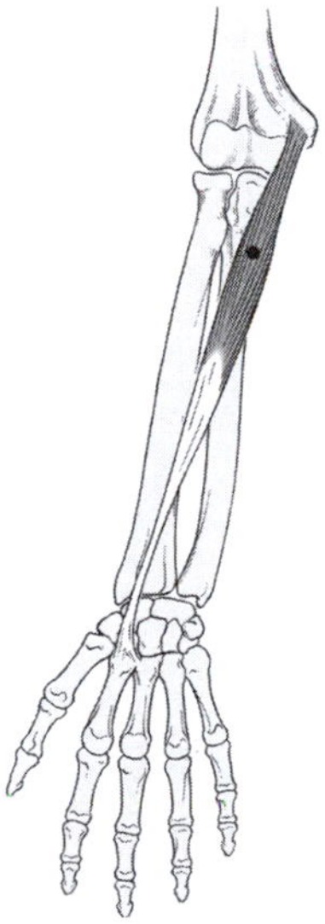

FIGURE 4.19

Muscle name ______________________

Origin ______________________

Insertion ______________________

Action ______________________

Innervation ______________________

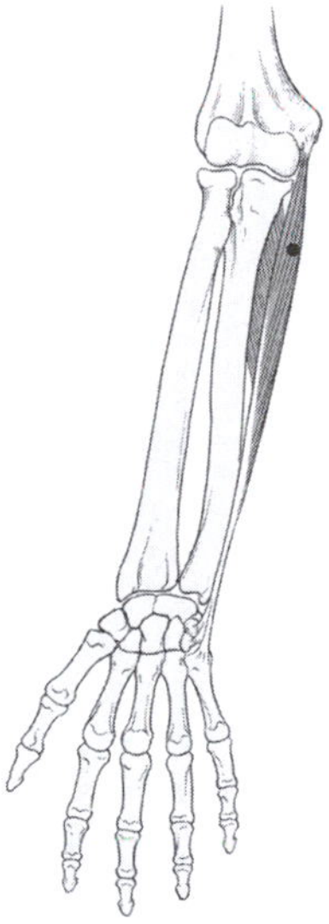

FIGURE 4.20

Muscle name ______________________

Origin ______________________

Insertion ______________________

Action ______________________

Innervation ______________________

FIGURE 4.18

Muscle name	Supinator
Origin (proximal attachment)	Lateral epicondyle of humerus, annular and radial collateral ligaments, and supinator crest of ulna
Insertion (distal attachment)	Lateral surface of the upper one-third of the body of the radius
Action	Supinates the forearm
Innervation	Radial nerve (C6)

FIGURE 4.17

Muscle name	Pronator Teres
Origin (proximal attachment)	**Humeral head**—just above the medial epicondyle of the humerus **Ulnar head**—medial side of the coronoid process of the ulna
Insertion (distal attachment)	Middle of lateral surface of radius
Action	Pronates the forearm and assists in flexing the elbow joint
Innervation	Median nerve (C6, C7)

FIGURE 4.20

Muscle name	Flexor Carpi Ulnaris
Origin (proximal attachment)	**Humeral head**—common tendon from the medial epicondyle of humerus **Ulnar head**—olecranon process and proximal two-thirds of the posterior border of the ulna
Insertion (distal attachment)	Pisiform bone, hook of the hamate, and the base of the fifth metacarpal bone
Action	Flexes and adducts the wrist
Innervation	Ulnar nerve (C7, C8)

FIGURE 4.19

Muscle name	Flexor Carpi Radialis
Origin (proximal attachment)	Medial epicondyle of the humerus
Insertion (distal attachment)	Base of second and third metacarpal bones
Action	Flexes wrist and abducts hand
Innervation	Median nerve (C6, C7)

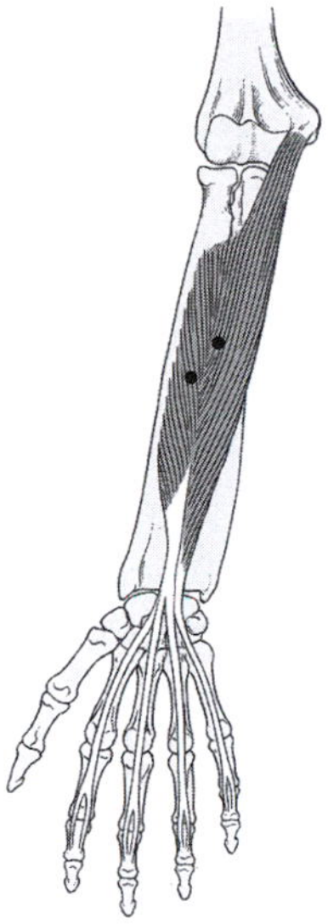

FIGURE 4.21

Muscle name ______________________

Origin ______________________

Insertion ______________________

Action ______________________

Innervation ______________________

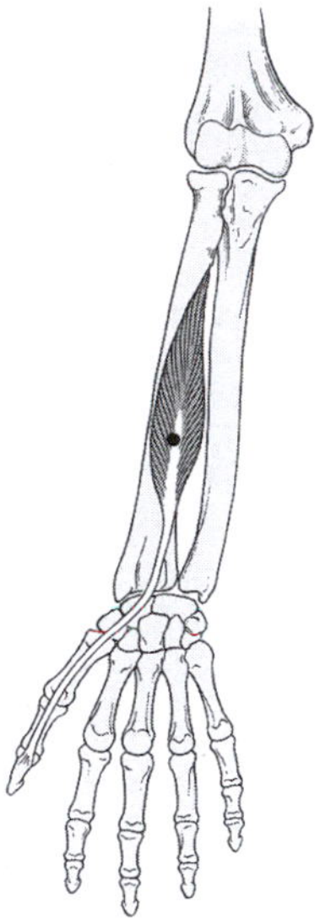

FIGURE 4.22

Muscle name ______________________

Origin ______________________

Insertion ______________________

Action ______________________

Innervation ______________________

FIGURE 4.23

Muscle name ______________________

Origin ______________________

Insertion ______________________

Action ______________________

Innervation ______________________

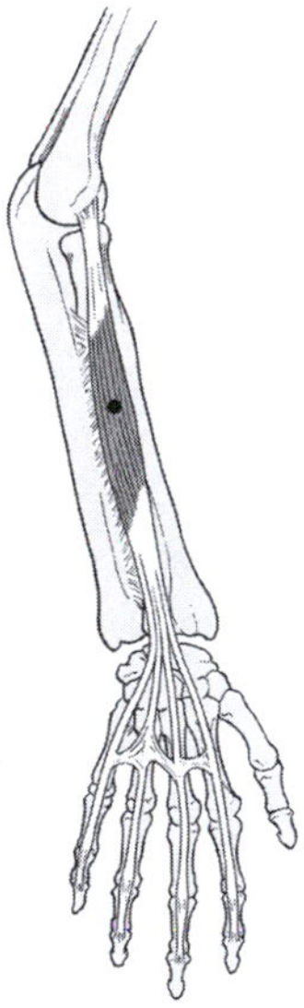

FIGURE 4.24

Muscle name ______________________

Origin ______________________

Insertion ______________________

Action ______________________

Innervation ______________________

FIGURE 4.21

Muscle name	Flexor Digitorum Superficialis
Origin (proximal attachment)	**Humeral head**—medial epicondyle of the humerus through the common tendon and the medial margin of the coronoid process of the ulna **Radial head**—anterior surface of the shaft of the radius
Insertion (distal attachment)	Four tendons divide into two slips each; the slips insert into the sides of the middle phalanges of the four fingers
Action	Flexes the wrist and the middle phalanges of fingers two through five
Innervation	Median nerve (C7–T1)

FIGURE 4.22

Muscle name	Flexor Pollicis Longus
Origin (proximal attachment)	Middle of anterior shaft of the radius and interosseous membrane
Insertion (distal attachment)	Palmar surface of the base of the distal phalanx of the thumb
Action	Flexes the thumb; assists in abduction of wrist
Innervation	Anterior interosseous branch of the median nerve (C8, T1)

FIGURE 4.23

Muscle name	Extensor Carpi Radialis Longus
Origin (proximal attachment)	Lower third of lateral supracondylar ridge of humerus
Insertion (distal attachment)	Dorsal surface of the base of the second metacarpal bone
Action	Extends the wrist and abducts the hand
Innervation	Radial nerve (C6, C7)

FIGURE 4.24

Muscle name	Extensor Digitorum
Origin (proximal attachment)	Common extensor tendon from the lateral epicondyle of humerus
Insertion (distal attachment)	By four tendons to the lateral and dorsal surfaces of all the phalanges of digits two through five
Action	Extends the fingers and the wrist
Innervation	Deep branch of radial nerve (C6–C8)

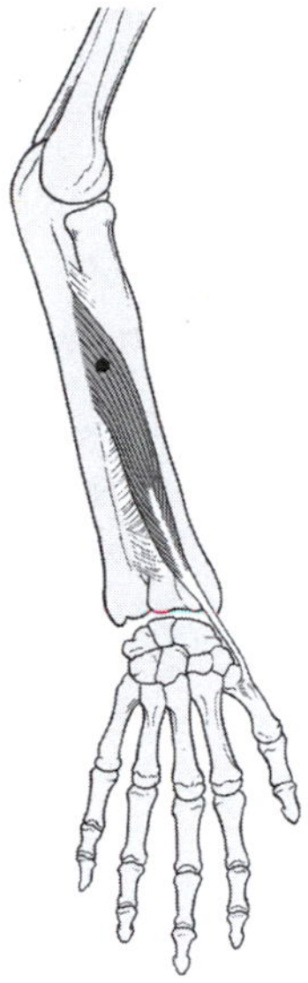

FIGURE 4.25

Muscle name ______________________________

Origin ______________________________

Insertion ______________________________

Action ______________________________

Innervation ______________________________

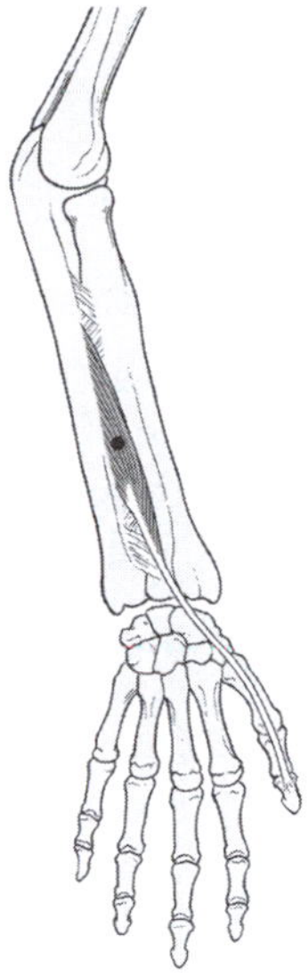

FIGURE 4.26

Muscle name ______________________________

Origin ______________________________

Insertion ______________________________

Action ______________________________

Innervation ______________________________

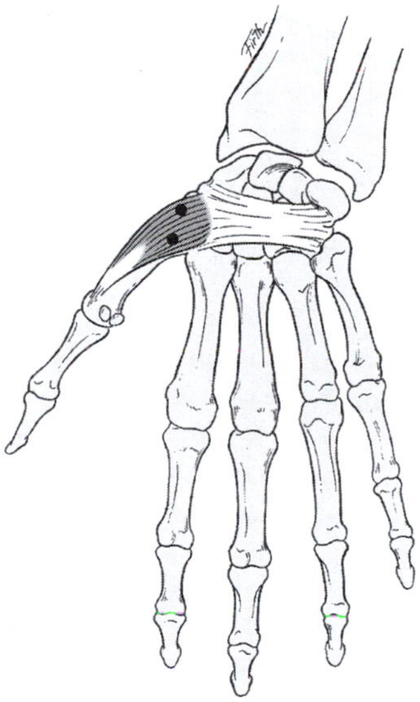

FIGURE 4.27

Muscle name ______________________________

Origin ______________________________

Insertion ______________________________

Action ______________________________

Innervation ______________________________

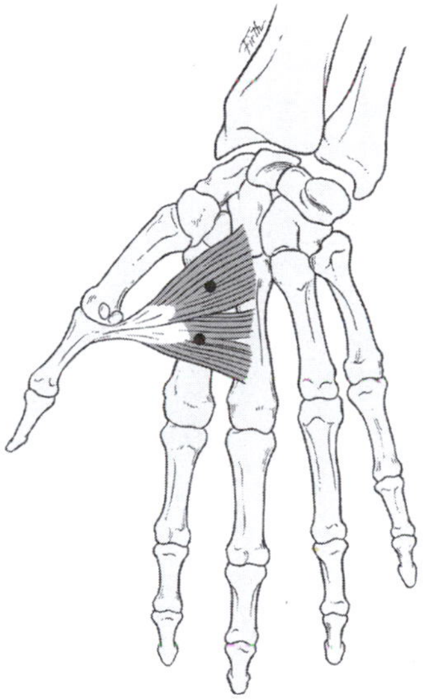

FIGURE 4.28

Muscle name ______________________________

Origin ______________________________

Insertion ______________________________

Action ______________________________

Innervation ______________________________

FIGURE 4.26

Muscle name	Extensor Pollicis Longus
Origin (proximal attachment)	Interosseous membrane and middle one-third of the posterior surface of the ulna
Insertion (distal attachment)	Dorsal surface of the base of the distal phalanx of the first digit
Action	Extends the interphalangeal joint and assists in extension of the metacarpophalangeal joints in the thumb; it also assists in abduction and extension of the wrist and the lateral rotation of the thumb
Innervation	Posterior interosseous branch of the radial nerve (C6–C8)

FIGURE 4.25

Muscle name	Abductor Pollicis Longus
Origin (proximal attachment)	Posterior surface of the body of the ulna distal to the origin of the supinator, the interosseous membrane, and the middle one-third of the body of the radius
Insertion (distal attachment)	Dorsal surface of the base of the first metacarpal bone
Action	Abducts and extends the first digit and abducts the wrist
Innervation	Posterior interosseous branch of the radial nerve (C6–C8)

FIGURE 4.28

Muscle name	Adductor Pollicis
Origin (proximal attachment)	**Oblique head**—anterior surfaces of second and third metacarpals, capitate and trapezoid bones **Transverse head**—anterior surface of third metacarpal bone
Insertion (distal attachment)	Medial side of base of proximal phalanx of thumb
Action	Adducts and flexes thumb
Innervation	Ulnar nerve (C8, T1)

FIGURE 4.27

Muscle name	Opponens Pollicis
Origin (proximal attachment)	Retinaculum (transverse carpal ligament) and trapezium bone
Insertion (distal attachment)	Anterior surface on the radial side of the first metacarpal bone
Action	Rotates thumb into opposition with fingers, acts together with other muscles of the thenar eminence to oppose thumb to fingers
Innervation	Median nerve (C7–T1)

II. MATCHING

Group A: In the blank to the left of each origin description, write the **LETTER** of the muscle with that origin.

1. _____ Lumbodorsal fascia, ribs, and iliac crest
2. _____ Ribs 1 through 8 or 9
3. _____ Supraspinous fossa of scapula
4. _____ Sternum, clavicle, and aponeurosis of external oblique muscles
5. _____ Spinous processes of T2 through T5
6. _____ Subscapular fossa of scapula
7. _____ Nuchal line, external occipital protuberance, and spinous processes of C1 through T12
8. _____ Spine of scapula, acromion process, and lateral one-third of clavicle
9. _____ Lower one-third of lateral border of scapula
10. _____ Transverse processes of C1 through C4

A. Deltoid
B. Trapezius
C. Latissimus dorsi
D. Teres major
E. Serratus anterior
F. Pectoralis major
G. Levator scapulae
H. Rhomboideus major
I. Subscapularis
J. Supraspinatus

Group B: In the blank to the left of each insertion description, write the **LETTER** of the muscle with that insertion.

1. _____ Posterior surface of olecranon process
2. _____ Middle of lateral surface of radius
3. _____ Coronoid process of ulna
4. _____ Upper one-third of lateral surface of radius
5. _____ Radial tuberosity and aponeurosis of flexor muscles
6. _____ Lateral surface of olecranon process and proximal surface of ulna
7. _____ Shaft of humerus opposite deltoid tuberosity
8. _____ Base of styloid process of radius

A. Coracobrachialis
B. Biceps brachii
C. Brachialis
D. Brachioradialis
E. Triceps brachii
F. Anconeus
G. Supinator
H. Pronator teres

Group C: In the blank to the left of each insertion description, write the **LETTER** of the muscle with that insertion.

1. _____ Pisiform, hook of hamate, and base of fifth metacarpal
2. _____ Dorsal surface of base of second metacarpal
3. _____ Dorsal surface of digits 2 through 5
4. _____ Base of second and third metacarpal bones
5. _____ Dorsal surface of fifth metacarpal
6. _____ Sides of middle phalanges of digits 2–5
7. _____ Flexor retinaculum
8. _____ Distal phalanx of thumb
9. _____ Dorsal surface of base of first metacarpal bone
10. _____ Dorsal surface of base of distal phalanx of fifth digit

A. Flexor carpi radialis
B. Palmaris longus
C. Flexor carpi ulnaris
D. Flexor pollicis longus
E. Flexor digitorum superficialis
F. Extensor carpi radialis longus
G. Extensor digitorum
H. Extensor carpi ulnaris
I. Extensor digiti minimi
J. Abductor pollicis longus

III. SENTENCE COMPLETION:

Circle the term in parentheses that correctly completes each statement.

1. All of the following are actions of the pectoralis major muscle *except* (**protraction of the scapula, abduction of the arm, medial rotation of the arm**).
2. The (**pectoralis minor, serratus anterior, subclavius, rhomboideus major**) muscle depresses the clavicle and draws the shoulder forward and downward.
3. The (**pectoralis minor, serratus anterior, subclavius, levator scapulae**) elevates the scapula.
4. The (**pectoralis major, trapezius, serratus anterior, rhomboideus major**) abducts the scapula.
5. The muscle important in bringing the arm down in a power stroke such as in hammering or rowing is the (**rhomboideus major, latissimus dorsi, deltoid, serratus anterior**).
6. The (**anterior, middle, posterior**) portion of the deltoid extends and laterally rotates the arm.
7. The rotator cuff muscles do *not* include the (**subscapularis, teres major, supraspinatus, infraspinatus**).
8. All of the following originate on the retinaculum *except* the (**adductor pollicis, opponens pollicis, flexor pollicis brevis, abductor pollicis brevis**).
9. The (**teres major, teres minor**) rotates the arm laterally.
10. The supraglenoid tubercle of the scapula is the origin of the (**long head, short head**) of the biceps brachii.
11. The flexors of the elbow include all *but* the (**biceps brachii, brachialis, anconeus, brachioradialis**).
12. The flexor digiti minimi is innervated by the (**musculocutaneus, ulnar, medial, radial**) nerve.
13. All of the following muscles flex the wrist *except* the (**flexor carpi radialis, palmaris longus, flexor pollicis longus, flexor digitorum superficialis**).
14. The muscle that flexes the middle phalanx of digits 2 through 5 is the (**flexor pollicis longus, flexor palmaris longus, flexor digitorum superficialis, flexor digitorum profundus**).

15. The (**pronator teres, pronator quadratus**) is located deep between the distal ends of the radius and ulna.

16. The anatomical snuff box is formed from the extensor pollicis brevis and the (**extensor digitorum, extensor pollicis longus, extensor indices, abductor pollicis longus**).

17. The (**abductor pollicis brevis, abductor pollicis longus, flexor pollicis brevis, extensor pollicis brevis**) assists in opposition of the thumb.

18. The rotator cuff muscle often implicated in frozen shoulder syndrome is the (**subscapularis, teres minor, supraspinatus, infraspinatus**).

19. The muscle that wrinkles the skin of the palm is the (**lumbricales, palmar interossei, palmaris brevis, extensor digitorum**).

20. All of the following abduct the hand at the wrist *except* the (**extensor carpi radialis brevis, extensor carpi ulnaris, abductor pollicis longus, flexor carpi radialis**).

IV. MULTIPLE CHOICE

In the blank to the left of each question, write the **LETTER** of the correct answer.

1. _____ A condition in which a person is missing the pectoralis major muscle on one side and also has some skeletal abnormalities is known as
   A. Dyspectoralis
   B. Kyphosis
   C. Poland's syndrome
   D. Bell's palsy

2. _____ The muscle that is important in pushing forward and punching is the
   A. Serratus anterior
   B. Trapezius
   C. Pectoralis major
   D. Rhomboideus major

3. _____ The muscle that works with the upper trapezius when shrugging the shoulder is the
   A. Rhomboideus major
   B. Levator scapulae
   C. Supraspinatus
   D. Latissimus dorsi

4. _____ The large muscle inserting in the floor of the intertubercular groove of the humerus is the
   A. Pectoralis minor
   B. Latissimus dorsi
   C. Deltoid
   D. Subscapularis

5. _____ All of the following are rotator cuff muscles *except* the
   A. Supraspinatus
   B. Rhomboideus minor
   C. Infraspinatus
   D. Teres minor

6. _____ All of the following muscles are adductors of the humerus *except* the
   A. Infraspinatus
   B. Teres major
   C. Pectoralis major
   D. Coracobrachialis

7. _____ The short head of the biceps brachii originates on the
   A. Acromion process of the scapula
   B. Lip of the glenoid cavity
   C. Lesser tubercle of the humerus
   D. Coracoid process of the scapula

8. _____ The muscle that stabilizes the elbow during rapid flexion and extension such as in hammering is the
   A. Pronator teres
   B. Brachioradialis
   C. Brachialis
   D. Biceps brachii

9. _____ The radial nerve often passes through the superficial and deep layers of what muscle?
   A. Supinator
   B. Flexor digitorum
   C. Anconeus
   D. Pronator teres

10. _____ The abductor pollicis brevis, opponens pollicis, and adductor pollicis all make up the
   A. Thenar eminence
   B. Midpalmar muscles
   C. Hypothenar eminence
   D. Snuff box

11. _____ The pectoralis major, coracobrachialis, latissimus dorsi, and teres major all work together to cause what type of movement of the glenohumeral joint?
   A. Extension
   B. Abduction
   C. Adduction
   D. Flexion

12. _____ Which of the following muscles does *not* extend the shoulder joint?
   A. Deltoid
   B. Teres major
   C. Triceps brachii
   D. Infraspinatus

13. _____ Which of the following muscles inserts on the olecranon process of the ulna?
   A. Brachioradialis
   B. Biceps brachii
   C. Triceps brachii
   D. Brachialis

14. _____ When rolling a bowling ball forward, the primary movement at the shoulder joint is
   A. Flexion
   B. Abduction
   C. Extension
   D. Circumduction

15. _____ The muscles that cause upward rotation of the scapula are the
   A. Lower trapezius and the pectoralis minor
   B. Upper trapezius and levator scapulae
   C. Upper trapezius and serratus anterior
   D. Serratus anterior and teres major

16. _____ Which of the following muscles helps to flex the glenohumeral joint?
   A. Brachioradialis
   B. Deltoid
   C. Infraspinatus
   D. Brachialis

17. _____ Which of the following muscles does *not* flex the wrist?
   A. Flexor carpi radialis longus
   B. Flexor carpi ulnaris
   C. Pronator quadratus
   D. Palmaris longus

18. _____ The primary flexor of the distal phalanges of the fingers is the
   A. Flexor carpi ulnaris
   B. Pollicis longus
   C. Flexor digitorum profundus
   D. Flexor digitorum superficialis

19. _____ Which of the following muscles is innervated by the axillary nerve?
   A. Deltoid
   B. Brachialis
   C. Pectoralis major
   D. Levator scapulae

20. _____ Which of the following muscles helps stabilize the scapula preparatory to movement?
   A. Serratus anterior and triceps brachii
   B. Rhomboideus muscles and gluteals
   C. Rhomboideus major and trapezius
   D. Pectoralis major and serratus anterior

V. FILL IN THE BLANK

Write the correct word(s) in the blank spaces to complete each statement.

1. The small muscle that depresses the clavicle and draws the shoulder forward and downward is the ____________________ .

2. The ____________________ muscle is the most important extensor of the forearm.

3. The muscle that flexes and abducts the hand at the wrist is the ____________________ .

4. The ____________________ muscle arises from the third through fifth rib and inserts on the coracoid process of the scapula.

5. The muscle that has been used as a cardiac assist muscle is the ____________________ muscle.

6. The muscle that originates on the spines of the vertebrae and inserts on the medial border of the scapula and that adducts, retracts, elevates, and rotates the scapula is the ____________________ .

7. The large muscle that arises from the lumbodorsal fascia and extends, adducts, and medially rotates the arm is the ____________________ .

8. The shoulder muscle that is one of the prime injection sites is the ____________________ .

9. The ____________________ and ____________________ muscles are located on the dorsal surface of the scapula.

10. Damage to the musculocutaneous nerve would limit ____________________ of the arm at the elbow.

11. The only flexor of the elbow that inserts on the styloid process of the radius is the ____________________ .

12. Movement to palms forward as in the anatomical position involves contraction of the ____________________ muscle.

13. The muscle that tenses the palmar fascia and flexes the hand at the wrist is the ____________________ .

14. The muscle that flexes the hand at the wrist and the middle digits of fingers 2 through 5 is the ____________________ .

15. The muscle that flexes the thumb is the ____________________ .

16. The muscle that abducts the thumb is the ____________________ .

17. The muscle that extends the little finger is the ___________________ .

18. The muscle that rotates the thumb into the opposition position is the ___________________ .

19. The group of four muscles that assist the extensor digitorum in extending the fingers are the ___________________ .

20. The deepest muscles of the hand are the ___________________ .

VI. SHORT ANSWER

Answer each of the questions in the space provided.

1. The sternalis muscles are found in only 1 of 20 people. In those in whom it does occur, it sometimes causes a misdiagnosis of a myocardial infarction. Why?

2. List the several muscles that work together to raise the scapula.

3. A patient complains of golfer's elbow. What causes this condition? Which specific tendon is strained?

4. A patient complains of carpel tunnel syndrome. Explain the condition to the patient. What commonly causes the condition? Which nerve is usually compressed?

5. What are the three primary actions of the muscles that originate from the medial supracondylar ridge and medial epicondyle of the humerus?

VII. CASE STUDIES

A. A 65-year-old woman was using 3-pound hand weights. After several repetitions, she heard a "popping" sound in her left shoulder. The next day, her shoulder was very painful, as was abducting or medially rotating the arm. She could not fasten her rear-buttoning blouse. After using ibuprofen and resting for a week, the shoulder was still painful. The orthopedic physician diagnosed the condition as arthritis in the A-C joint and impingement syndrome.

1. What bony features make up the A-C joint?

2. What muscle is being impinged?

3. What skeletal feature is it impinging on?

4. What is the primary action of the impinging muscle?

5. What other muscles must have also been strained, causing difficulty in fastening the blouse?

B. A 35-year-old secretary complains of lingering pain and weakness in her right hand, especially the thumb and first two fingers, with the loss of ability to grasp. She has impaired perception of touch to the thumb and first two fingers.

1. What is one probable diagnosis?

2. Which flexor muscles pass under the flexor retinaculum?

3. Which other two thumb muscles would be affected?

4. Which thumb motions would be compromised?

5. Which nerves are being compressed?

Muscles of the Hip, Thigh, and Lower Leg

5

At the end of this chapter, you should be able to

1. Identify the muscles of the hip and leg.
2. Name the points of attachment of the major hip and leg muscles.
3. Name the actions of the major hip and leg muscles.
4. Identify the three hamstring muscles.
5. Identify the four parts of the quadriceps muscle group.
6. Identify the muscles in the anterior compartment of the lower leg.
7. Identify the muscles in the posterior compartment of the lower leg.
8. Explain what a "high pointer" injury is.
9. Explain what a "pulled groin" injury is.
10. Identify the two muscles that insert together by the Achilles tendon.
11. Name the muscles in the first layer of the sole of the foot.
12. Name the muscles in the second layer of the sole of the foot.
13. Name the muscles in the third layer of the sole of the foot.

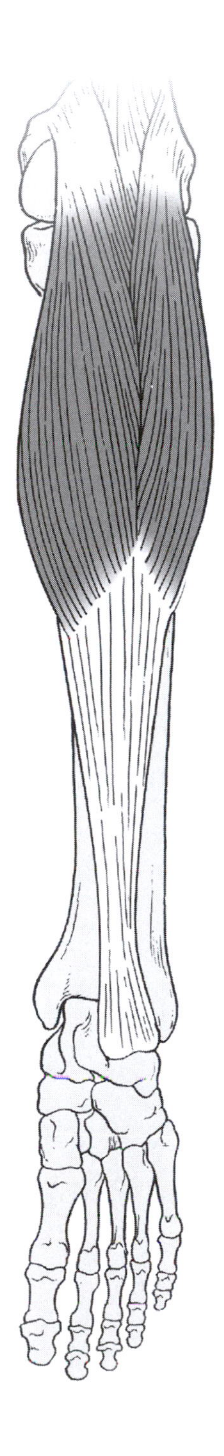

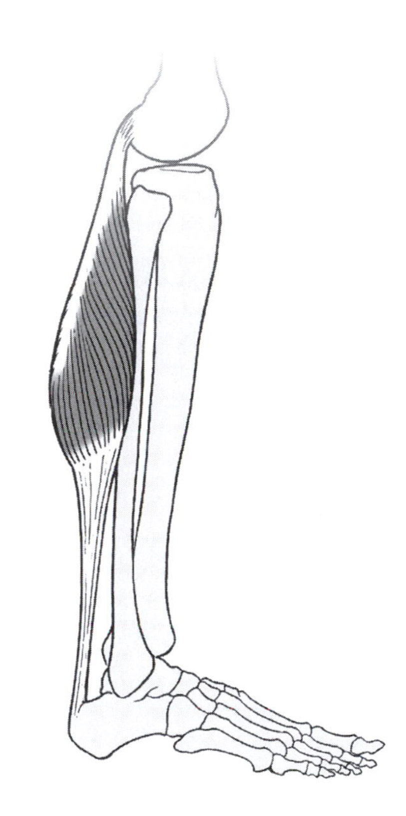

Word Search Puzzle

INSTRUCTIONS: Find and CIRCLE each of the listed terms in the Word Search Puzzle. (Terms may read from left to right, right to left, up, down, or diagonally.)

S U S O N A R B M E M I M E S X Z P V S

P S A O S P X F C B R S Q S U C A I L I

Q U A D R A T U S F X I V W Y K Z R Z R

T E N S O R F A C I A L A T A E Q I Z O

I S P O P L I T E U S I R O X E L F V M

B U P P Z Q B X S P E C I B X F D O C E

I T F E Z X U Y U Q B A C C S X Y R Q F

A S X C Q F L K L H D R D U I F G M H S

L A P T X D A L L G B G E D C X Q I J U

I V Q I Z C R X E Z T T H G U Z D S X T

S F S N B X I O M X U S W W L C J Z Q C

G S U E L O S Z E L R Z Y V L J T X J E

B D C U A B G H G Y P X X Q A X Z O D R

X Y Z S A R T O R I U S H G H Q J A R Z

ADDUCTOR
BICEPS
FLEXOR
FIBULARIS
GEMELLUS
GLUTEUS
GRACILIS
HALLUCIS
ILIACUS
PECTINEUS
PIRIFORMIS

POPLITEUS
PSOAS
QUADRATUS
RECTUS FEMORIS
SARTORIUS
SEMIMEMBRANOSUS
SOLEUS
TENSOR FACIA LATAE
TIBIALIS
VASTUS

I. ILLUSTRATION IDENTIFICATION (pages 95–108 can be cut and used as flash cards)

Write the technical information for each muscle in the spaces beneath the illustrations.

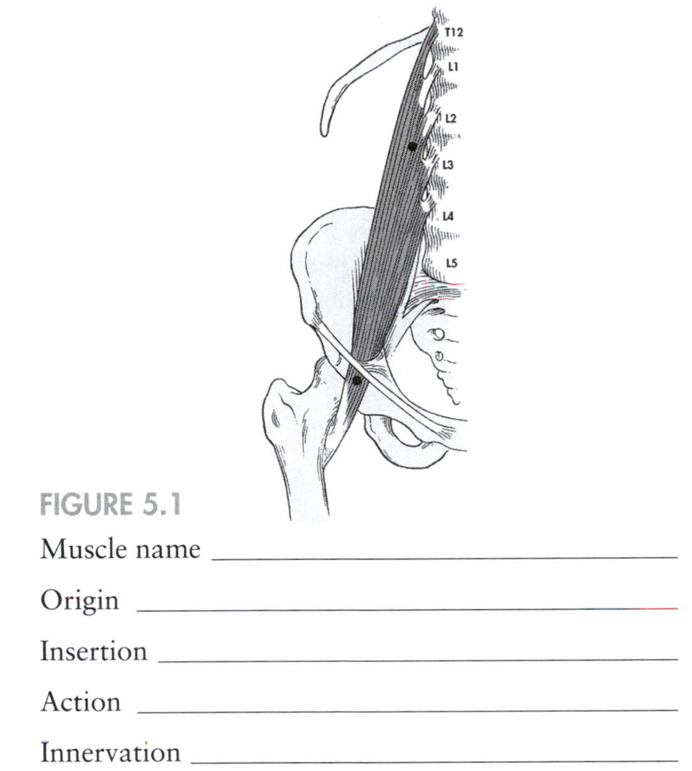

FIGURE 5.1

Muscle name ______________________

Origin ______________________

Insertion ______________________

Action ______________________

Innervation ______________________

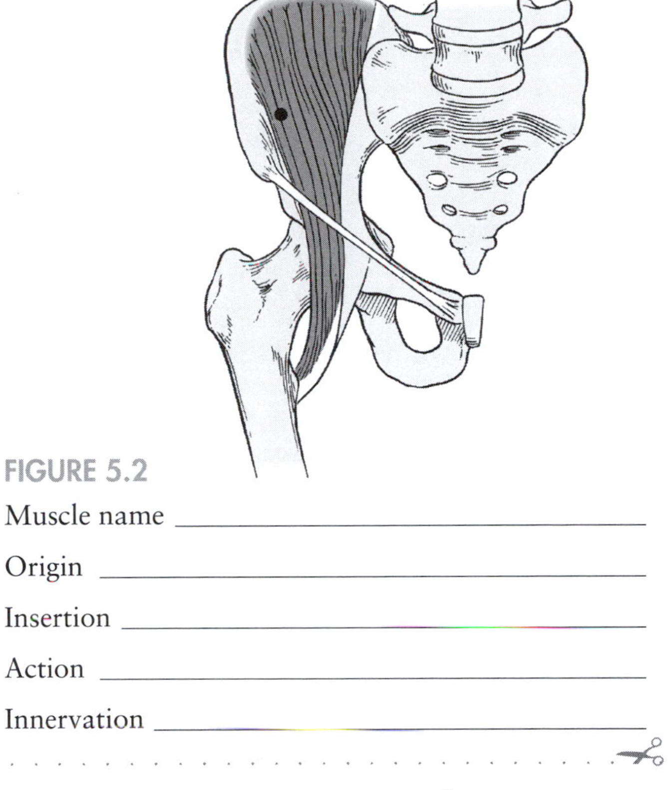

FIGURE 5.2

Muscle name ______________________

Origin ______________________

Insertion ______________________

Action ______________________

Innervation ______________________

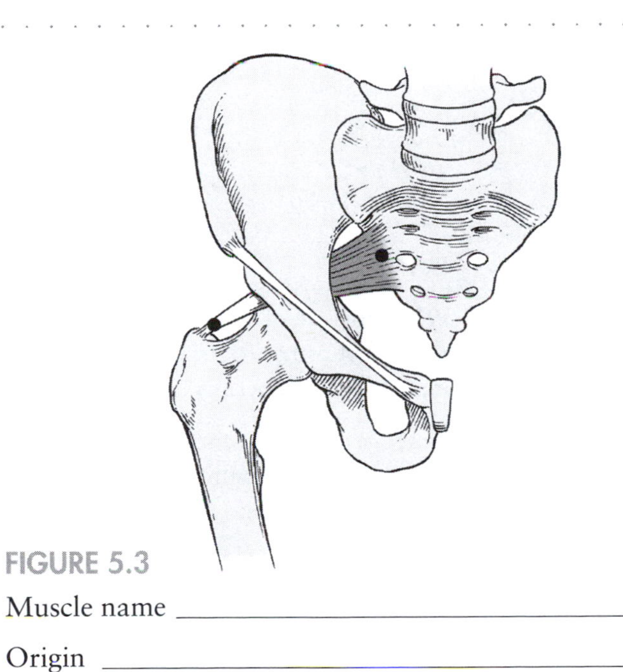

FIGURE 5.3

Muscle name ______________________

Origin ______________________

Insertion ______________________

Action ______________________

Innervation ______________________

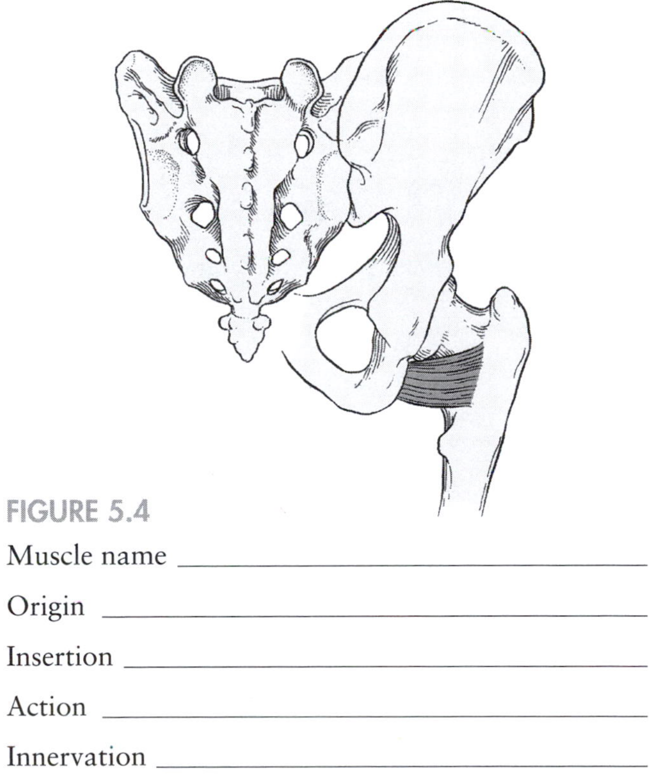

FIGURE 5.4

Muscle name ______________________

Origin ______________________

Insertion ______________________

Action ______________________

Innervation ______________________

FIGURE 5.2

Muscle name	Iliacus
Origin (superior attachment)	Upper two-thirds of iliac fossa, ala of the sacrum, anterior inferior iliac spine, and iliac crest
Insertion (inferior attachment)	With psoas major, lesser trochanter of femur
Action	Flexes thigh at hip joint
Innervation	Muscular branches of femoral nerve (L2–L4)

FIGURE 5.1

Muscle name	Psoas Major
Origin (superior attachment)	Transverse processes of all lumbar vertebrae, bodies of last thoracic and all lumbar vertebrae, and intervertebral disk of each lumbar vertebrae
Insertion (inferior attachment)	Lesser trochanter of femur
Action	Flexes thigh at the hip joint and flexes vertebral column
Innervation	Ventral rami of L2–L4

FIGURE 5.4

Muscle name	Quadratus Femoris
Origin (proximal attachment)	Upper part of the lateral border of the ischial tuberosity
Insertion (distal attachment)	Trochanteric crest of femur
Action	Laterally rotates the thigh at the hip joint
Innervation	Branch from sacral plexus (L5, S1)

FIGURE 5.3

Muscle name	Piriformis
Origin (proximal attachment)	Pelvic surface of the sacrum between the first through fourth sacral foramina and sacrotuberous ligament
Insertion (distal attachment)	Superior border of the greater trochanter of the femur
Action	Laterally rotates thigh at the hip joint and abducts thigh
Innervation	Anterior rami of S1, S2

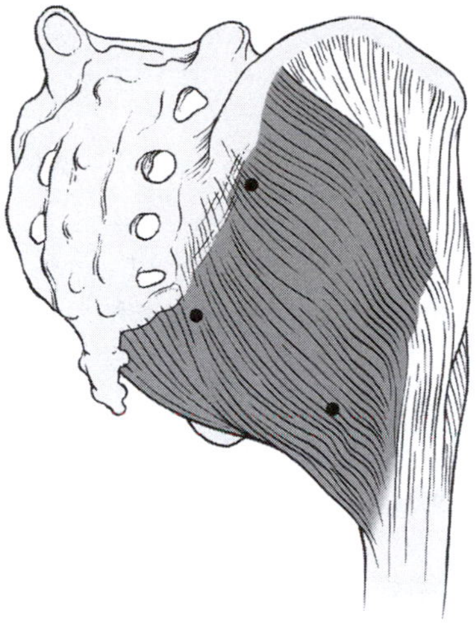

FIGURE 5.5

Muscle name ______________________

Origin ______________________

Insertion ______________________

Action ______________________

Innervation ______________________

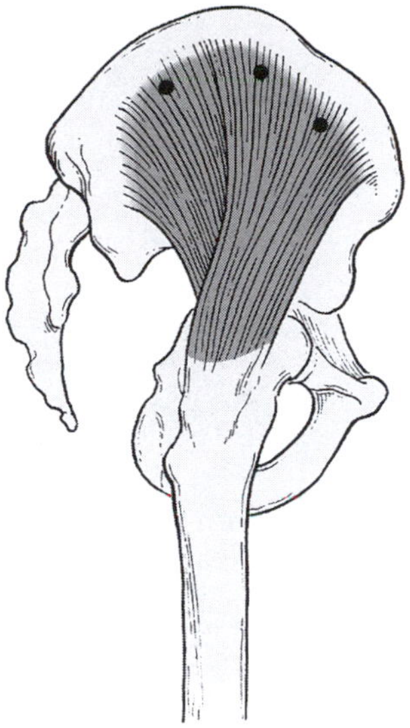

FIGURE 5.6

Muscle name ______________________

Origin ______________________

Insertion ______________________

Action ______________________

Innervation ______________________

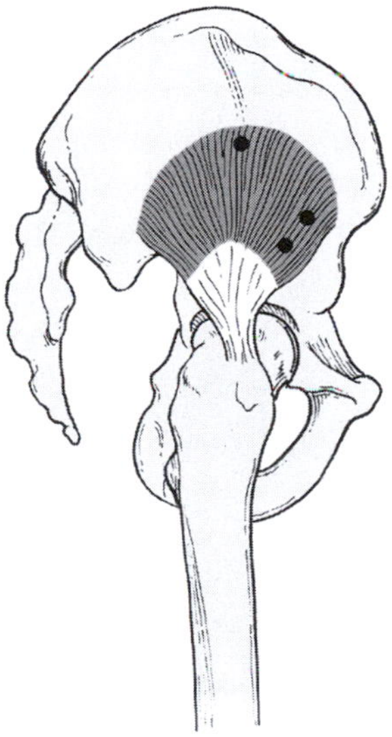

FIGURE 5.7

Muscle name ______________________

Origin ______________________

Insertion ______________________

Action ______________________

Innervation ______________________

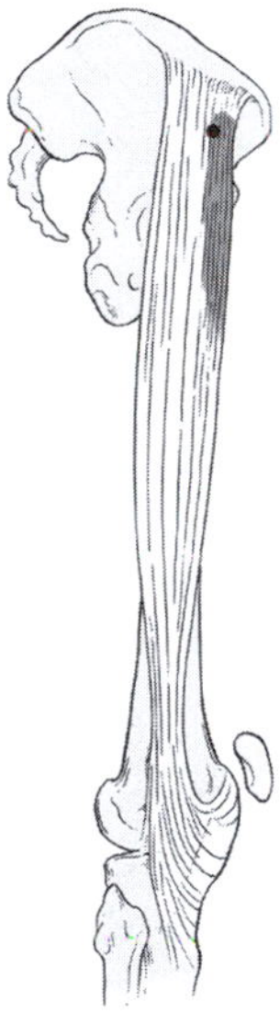

FIGURE 5.8

Muscle name ______________________

Origin ______________________

Insertion ______________________

Action ______________________

Innervation ______________________

FIGURE 5.5

Muscle name	Gluteus Maximus
Origin (proximal attachment)	Posterior gluteal line of ilium, adjacent posterior surface of sacrum and coccyx, sacrotuberous ligament, and aponeurosis of erector spinae muscles
Insertion (distal attachment)	Iliotibial tract of fascia lata and gluteal tuberosity of femur
Action	**Upper part**—extends and laterally rotates thigh **Lower part**—extends, laterally rotates thigh and assists in raising the trunk from a flexed position; also assists in adduction of the hip joint
Innervation	Inferior gluteal nerve (L5–S2)

FIGURE 5.6

Muscle name	Gluteus Medius
Origin (proximal attachment)	Between anterior and posterior gluteal lines on lateral surface of ilium
Insertion (distal attachment)	Lateral surface of the greater trochanter of the femur
Action	Abducts the hip joint and medially rotates the thigh; extends and flexes hip
Innervation	Superior gluteal nerve (L5, S1)

FIGURE 5.7

Muscle name	Gluteus Minimus
Origin (proximal attachment)	Outer surface of the ilium between the middle and inferior gluteal lines
Insertion (distal attachment)	Anterior border of the greater trochanter
Action	Abducts the femur at the hip joint and medially rotates the thigh; flexes hip
Innervation	Superior gluteal nerve (L4–S1)

FIGURE 5.8

Muscle name	Tensor Fasciae Latae
Origin (proximal attachment)	Anterior aspect of the outer lip of the iliac crest and the anterior superior iliac spine
Insertion (distal attachment)	Middle and proximal thirds of the thigh along the iliotibial tract. The iliotibial band inserts on the lateral epicondyle of tibia
Action	Assists in abduction, medial rotation, and flexion of thigh. Makes the iliotibial tract taut. Stabilizer of the hip
Innervation	Superior gluteal nerve (L4–S1)

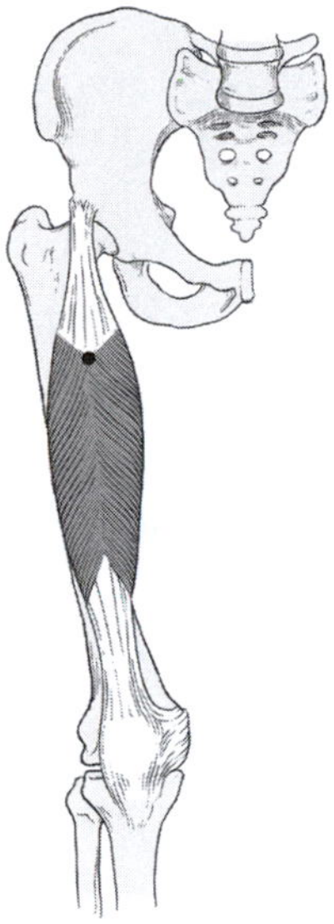

FIGURE 5.9

Muscle name ______

Origin ______

Insertion ______

Action ______

Innervation ______

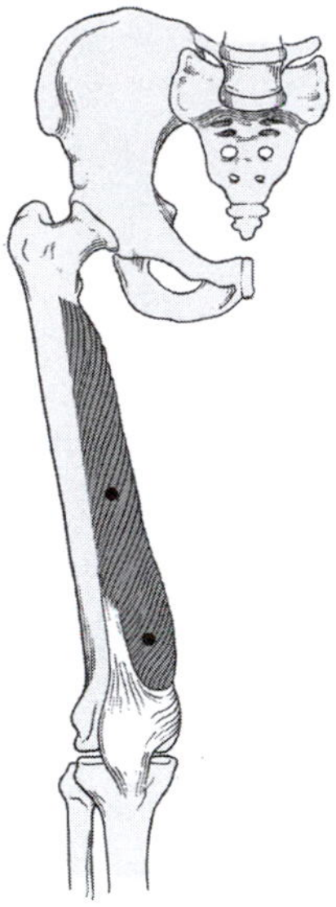

FIGURE 5.10

Muscle name ______

Origin ______

Insertion ______

Action ______

Innervation ______

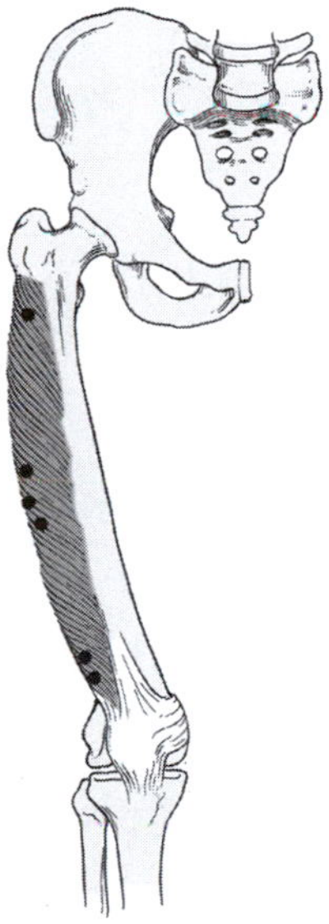

FIGURE 5.11

Muscle name ______

Origin ______

Insertion ______

Action ______

Innervation ______

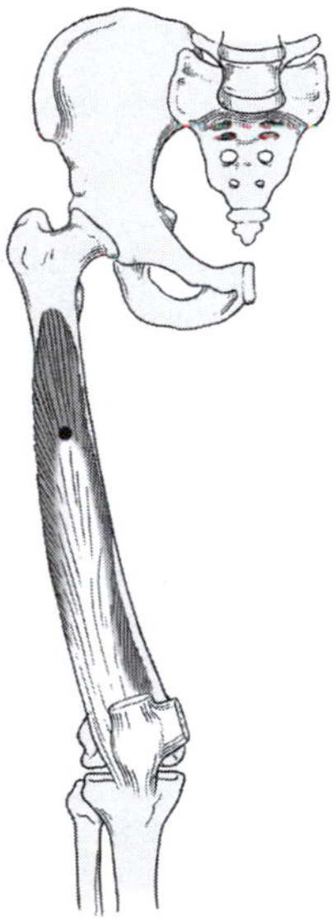

FIGURE 5.12

Muscle name ______

Origin ______

Insertion ______

Action ______

Innervation ______

FIGURE 5.10

Muscle name	Vastus Medialis
Origin (proximal attachment)	Lower half of intertrochanteric line, linea aspera, medial supracondylar line, and medial intermuscular septum
Insertion (distal attachment)	Medial border of the patella and then by the patella ligament to the tibial tuberosity
Action	Extends the leg at the knee joint and draws the patella medially
Innervation	Femoral nerve (L2–L4)

FIGURE 5.9

Muscle name	Rectus Femoris
Origin (proximal attachment)	**Anterior head**—anterior inferior iliac spine **Posterior head**—upper margin of the acetabulum
Insertion (distal attachment)	Patella and by the patella ligament to the tibial tuberosity
Action	Extends the leg at the knee and flexes the thigh at the hip joint
Innervation	Femoral nerve (L2–L4)

FIGURE 5.12

Muscle name	Vastus Intermedius
Origin (proximal attachment)	Anterior and lateral surfaces of the proximal two-thirds of the body of the femur
Insertion (distal attachment)	Deep surface of the tendon of the rectus femoris and vastus muscles; patella and through the patella ligament to the tibial tuberosity
Action	Extends the knee at the joint
Innervation	Femoral nerve (L2–L4)

FIGURE 5.11

Muscle name	Vastus Lateralis
Origin (proximal attachment)	Proximal intertrochanteric line, greater trochanter, gluteal tuberosity, and linea aspera
Insertion (distal attachment)	Patella and through the patella ligament to the tibial tuberosity
Action	Extends the knee joint and exerts a lateral pull on the patella
Innervation	Femoral nerve (L2–L4)

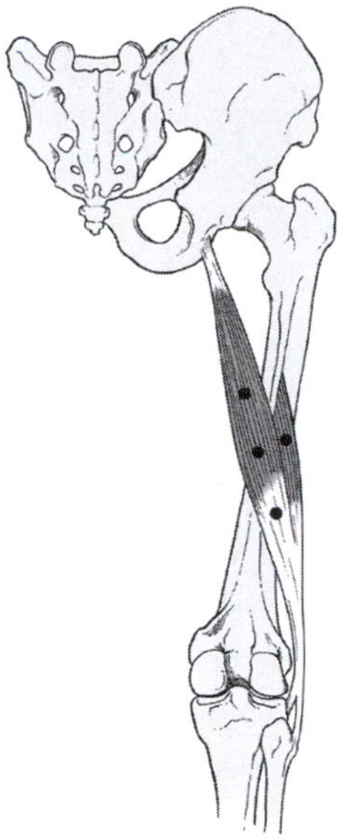

FIGURE 5.13

Muscle name ____________________

Origin ____________________

Insertion ____________________

Action ____________________

Innervation ____________________

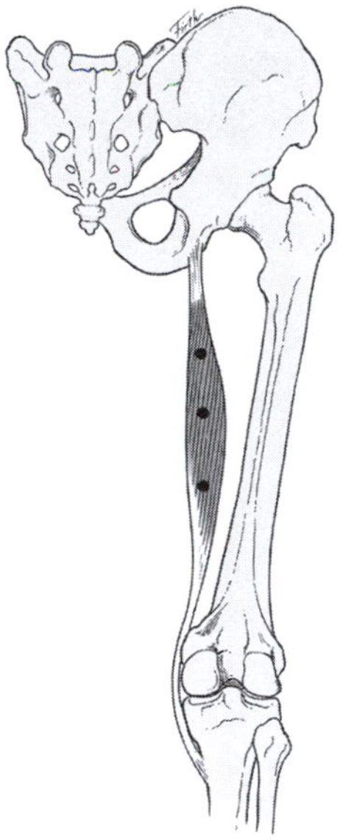

FIGURE 5.14

Muscle name ____________________

Origin ____________________

Insertion ____________________

Action ____________________

Innervation ____________________

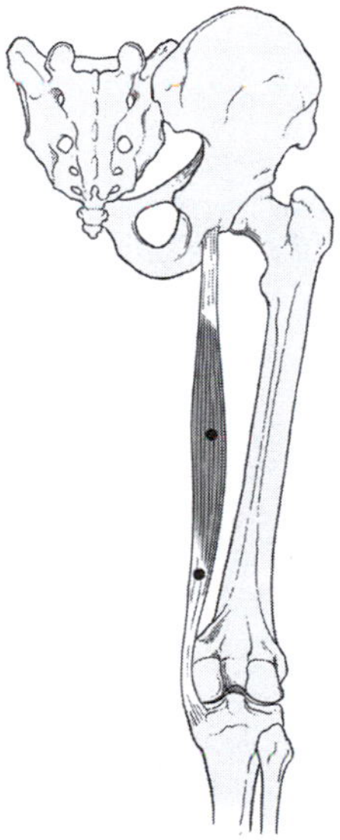

FIGURE 5.15

Muscle name ____________________

Origin ____________________

Insertion ____________________

Action ____________________

Innervation ____________________

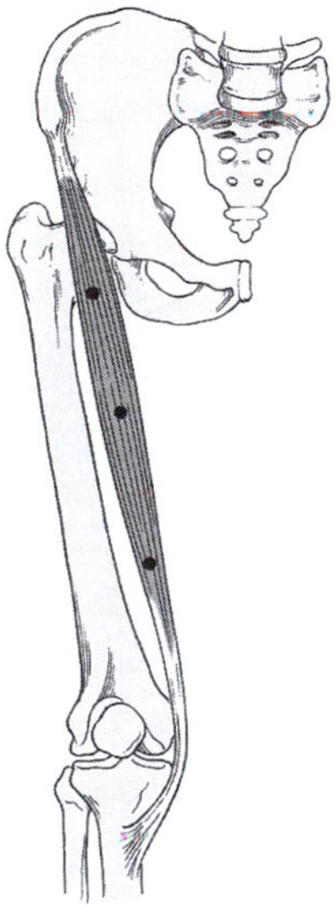

FIGURE 5.16

Muscle name ____________________

Origin ____________________

Insertion ____________________

Action ____________________

Innervation ____________________

FIGURE 5.13

Muscle name	Biceps Femoris
Origin (proximal attachment)	**Long head**—ischial tuberosity **Short head**—lateral lip of linea aspera, proximal two-thirds of supracondylar line
Insertion (distal attachment)	Common tendon passes downward to insert on head of fibula and lateral condyle of the tibia
Action	Flexes and laterally rotates the knee joint and extends the thigh
Innervation	**Long head**—tibial division of the sciatic nerve (L5–S2) **Short head**—fibular division of the sciatic nerve (L5–S2)

FIGURE 5.14

Muscle name	Semitendinosus
Origin (proximal attachment)	Ischial tuberosity
Insertion (distal attachment)	Upper medial surface of the shaft of the tibia
Action	Flexes and slightly medially rotates leg at knee joint, and extends the thigh at the hip joint
Innervation	Tibial portion of sciatic nerve (L5–S2)

FIGURE 5.15

Muscle name	Semimembranosus
Origin (proximal attachment)	Ischial tuberosity
Insertion (distal attachment)	Posterior part of the medial condyle of tibia
Action	Flexes and slightly medially rotates leg at knee joint and extends thigh at hip
Innervation	Tibial portion of sciatic nerve (L5–S2)

FIGURE 5.16

Muscle name	Sartorius
Origin (proximal attachment)	Anterior superior iliac spine and upper half of iliac notch
Insertion (distal attachment)	Proximal part of the medial aspect of the tibia
Action	Flexes, laterally rotates, and abducts the hip joint; also flexes the torso toward the leg, and flexes and assists in medial rotation of the knee
Innervation	Femoral nerve (L2, L3)

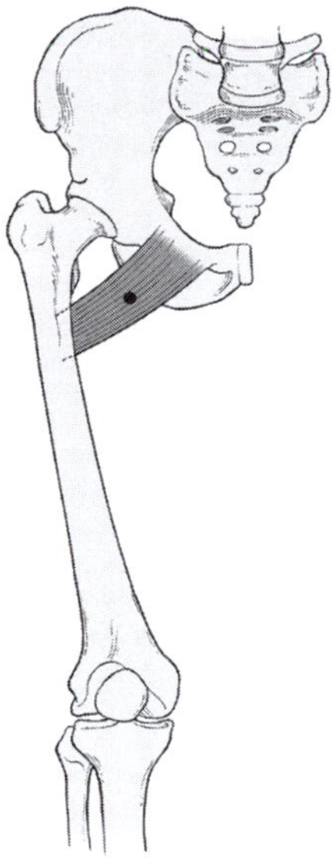

FIGURE 5.17

Muscle name ______________________

Origin ______________________

Insertion ______________________

Action ______________________

Innervation ______________________

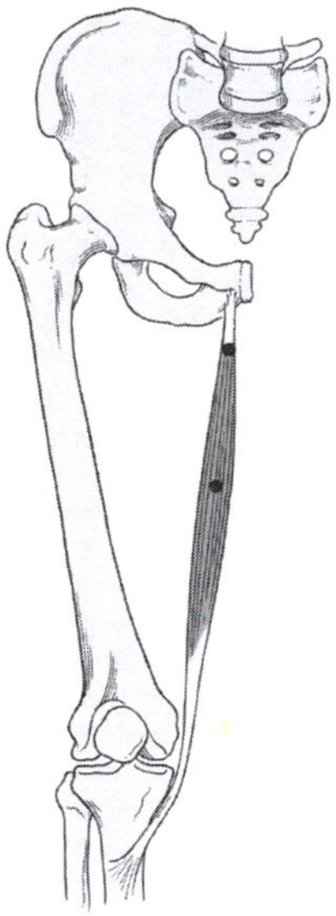

FIGURE 5.18

Muscle name ______________________

Origin ______________________

Insertion ______________________

Action ______________________

Innervation ______________________

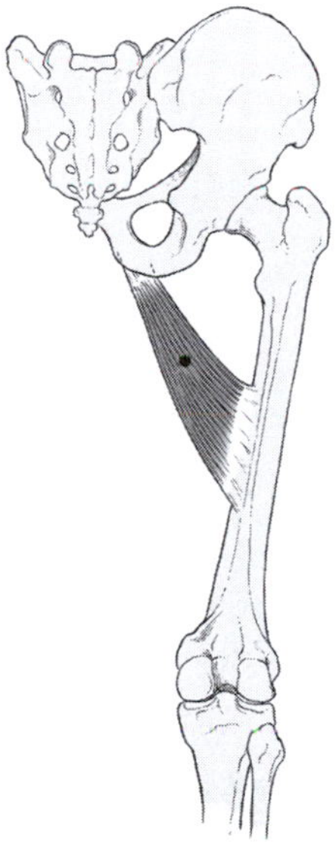

FIGURE 5.19

Muscle name ______________________

Origin ______________________

Insertion ______________________

Action ______________________

Innervation ______________________

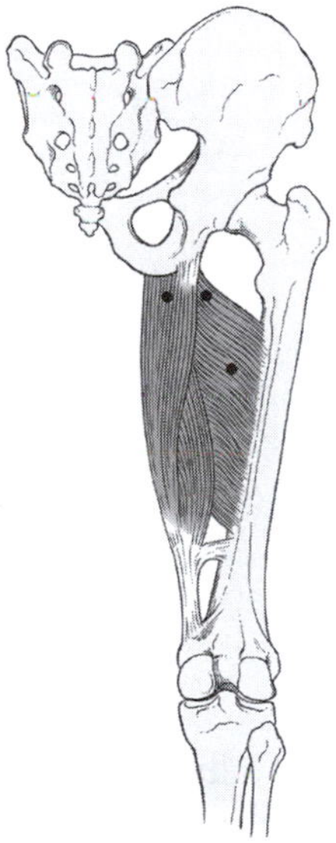

FIGURE 5.20

Muscle name ______________________

Origin ______________________

Insertion ______________________

Action ______________________

Innervation ______________________

FIGURE 5.18

Muscle name	Gracilis
Origin (proximal attachment)	Inferior ramus and body of pubis
Insertion (distal attachment)	Medial surface of tibia just inferior to its medial condyle
Action	Adducts thigh at hip joint and flexes leg at knee joint; assists in medial rotation and flexes hip
Innervation	Obturator nerve (L3, L4)

FIGURE 5.17

Muscle name	Pectineus
Origin (proximal attachment)	Pectineal line on superior ramus of pubis
Insertion (distal attachment)	From lesser trochanter to linea aspera of femur
Action	Flexes femur at hip and assists in adduction of femur at hip
Innervation	Femoral nerve (L2–L4)

FIGURE 5.20

Muscle name	Adductor Magnus
Origin (proximal attachment)	Inferior ramus of pubis and ramus of ischium and inferior portion of ischial tuberosity
Insertion (distal attachment)	Linea aspera and adductor tubercle of femur
Action	Adducts and extends thigh; assists in medial rotation
Innervation	Obturator and sciatic nerves (L2–L4)

FIGURE 5.19

Muscle name	Adductor Longus
Origin (proximal attachment)	Anterior body of pubis
Insertion (distal attachment)	Medial one-third of medial lip of linea aspera of femur
Action	Adducts and flexes thigh; assists in medial rotation
Innervation	Obturator nerve (L2–L4)

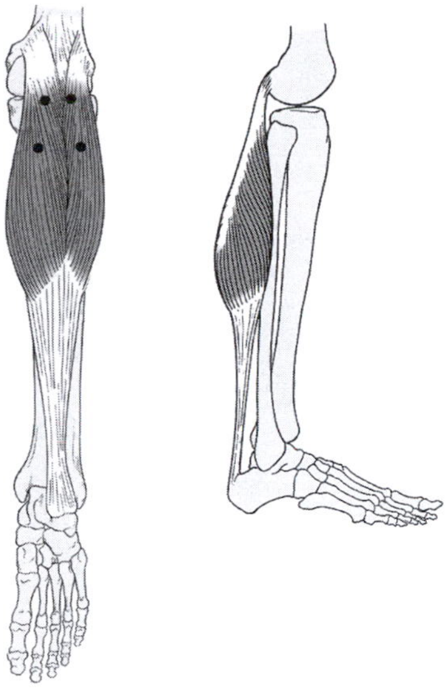

FIGURE 5.21

Muscle name ____________________

Origin ____________________

Insertion ____________________

Action ____________________

Innervation ____________________

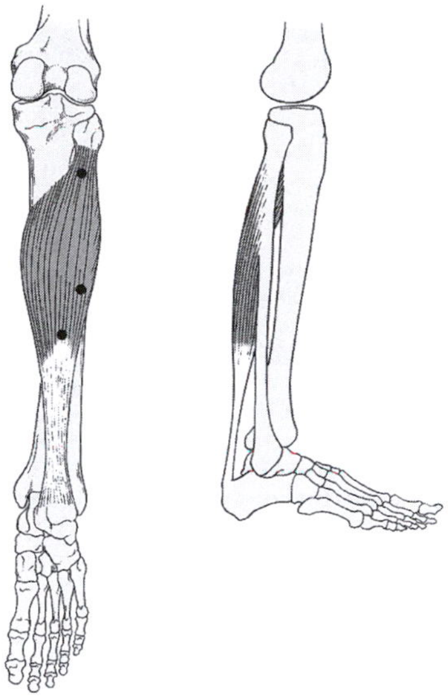

FIGURE 5.22

Muscle name ____________________

Origin ____________________

Insertion ____________________

Action ____________________

Innervation ____________________

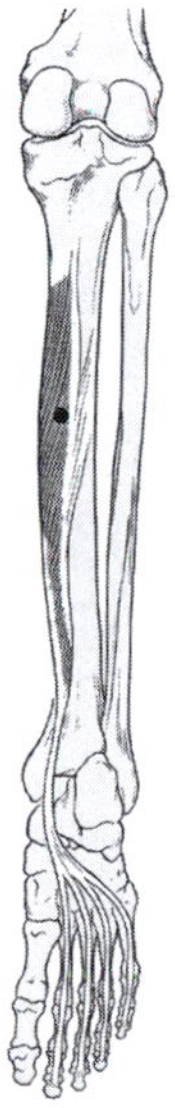

FIGURE 5.23

Muscle name ____________________

Origin ____________________

Insertion ____________________

Action ____________________

Innervation ____________________

FIGURE 5.24

Muscle name ____________________

Origin ____________________

Insertion ____________________

Action ____________________

Innervation ____________________

FIGURE 5.22

Muscle name	Soleus
Origin (proximal attachment)	Upper one-fourth of posterior surface of the fibula, soleal line, and upper shaft of tibia
Insertion (distal attachment)	With the gastrocnemius, via the Achilles tendon to the calcaneous
Action	Plantar flexion of the ankle and stabilizes the leg over the foot
Innervation	Tibial nerve (S1, S2)

FIGURE 5.21

Muscle name	Gastrocnemius
Origin (proximal attachment)	**Medial head**—upper posterior part of medial condyle of femur **Lateral head**—supracondylar line and lateral condyle of the femur
Insertion (distal attachment)	Calcaneous via the Achilles tendon
Action	Plantar flexion of the ankle joint and assists in flexion of the knee joint
Innervation	Tibial nerve (S1, S2)

FIGURE 5.24

Muscle name	Tibialis Posterior
Origin (proximal attachment)	Lateral part of posterior surface of tibia, interosseous membrane, and proximal half of posterior surface of fibula
Insertion (distal attachment)	Tuberosity of navicular bone, cuboid, cuneiforms, second, third, and fourth metatarsals and calcaneus
Action	Plantar flexion and inversion of the foot
Innervation	Tibial nerve (L5, S1)

FIGURE 5.23

Muscle name	Flexor Digitorum Longus
Origin (proximal attachment)	Medial part of posterior surface of tibia
Insertion (distal attachment)	Plantar surface of the bases of the distal phalanges of the second, third, fourth, and fifth toes
Action	Flexes distal phalanges of lateral four toes, assists in plantar flexion of foot, inverts foot
Innervation	Tibial nerve (L5, S1)

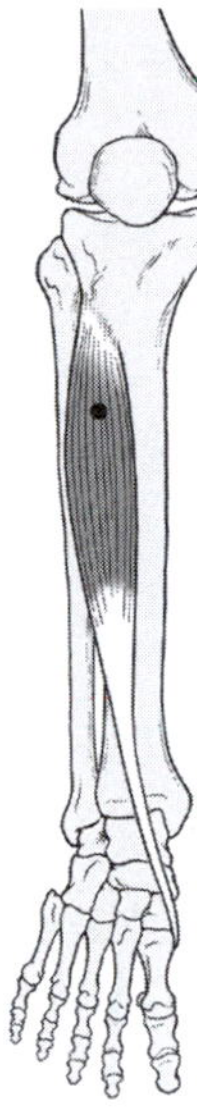

FIGURE 5.25

Muscle name ______________________

Origin ______________________

Insertion ______________________

Action ______________________

Innervation ______________________

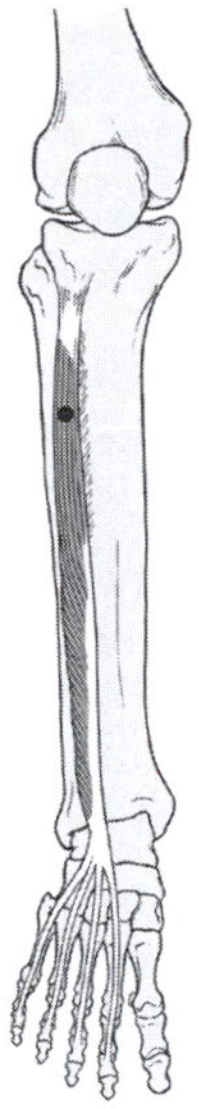

FIGURE 5.26

Muscle name ______________________

Origin ______________________

Insertion ______________________

Action ______________________

Innervation ______________________

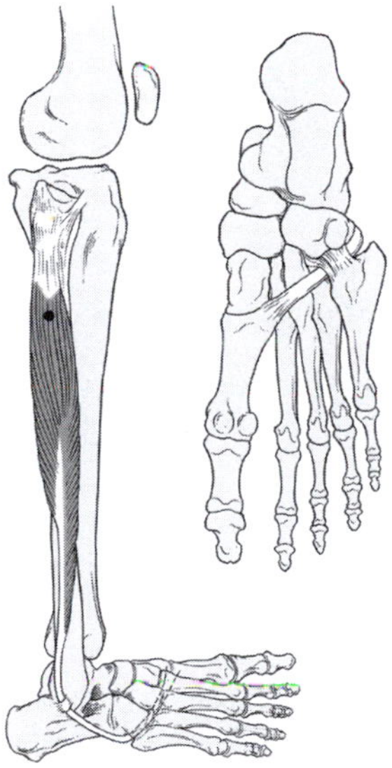

FIGURE 5.27

Muscle name ______________________

Origin ______________________

Insertion ______________________

Action ______________________

Innervation ______________________

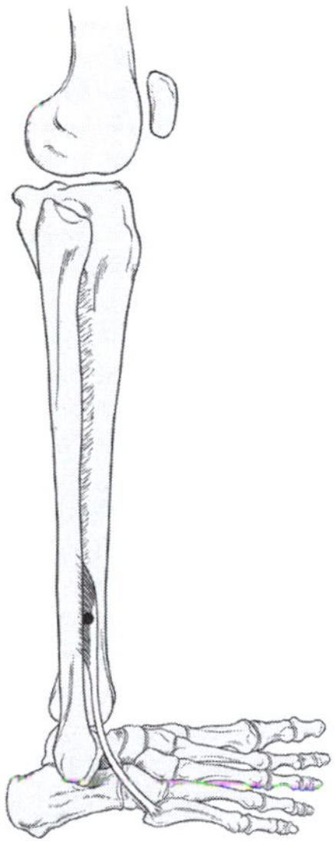

FIGURE 5.28

Muscle name ______________________

Origin ______________________

Insertion ______________________

Action ______________________

Innervation ______________________

FIGURE 5.26

Muscle name	Extensor Digitorum Longus
Origin (proximal attachment)	Lateral condyle of the tibia, proximal three-fourths of the anterior surface of the fibula, and the interosseous membrane
Insertion (distal attachment)	By four tendons to the second through fifth digits. Each tendon divides into a middle slip that inserts in the base of the middle phalanx and two lateral slips that insert in the base of the distal phalanx
Action	Extends the phalanges of the second through fifth digits, assists in dorsiflexion of the ankle and in eversion of the foot
Innervation	Deep fibular nerve (L4–S1)

FIGURE 5.25

Muscle name	Tibialis Anterior
Origin (proximal attachment)	Lateral condyle and proximal one-half of the lateral surface of the tibia and the interosseous membrane
Insertion (distal attachment)	Medial plantar surface of the first cuneiform bone and the base of the first metatarsal bone
Action	Dorsiflexion of the foot at the ankle joint and inversion of the foot
Innervation	Deep fibular nerve (L4–S1)

FIGURE 5.28

Muscle name	Fibularis Tertius—Peroneus Tertius
Origin (proximal attachment)	Lower third of anterior surface of the fibula and the interosseous membrane
Insertion (distal attachment)	Dorsal surface of the base of the fifth metatarsal bone
Action	Dorsiflexion and eversion of the foot
Innervation	Deep fibular nerve (L4–S1)

FIGURE 5.27

Muscle name	Fibularis Longus—Peroneus Longus
Origin (proximal attachment)	Upper two-thirds of lateral surface of the fibula
Insertion (distal attachment)	Lateral side of medial cuneiform and the base of the first metatarsal
Action	Plantar flexion and eversion of the foot
Innervation	Superficial fibular nerve (L4–S1)

II. MATCHING

Group A: In the blank to the left of each origin description, write the **LETTER** of the muscle with that origin.

1. _____ Posterior gluteal line of ilium, posterior surface of sacrum and coccyx
2. _____ Anterior aspect of outer lip of iliac crest and anterior superior iliac spine
3. _____ Transverse processes and bodies of lumbar vertebrae and intervertebral disks
4. _____ Lateral surface of ilium between anterior and posterior gluteal lines
5. _____ Outer surface of superior and inferior rami of pubis and ramus of ischium

A. Psoas major
B. Obturator externus
C. Gluteus maximus
D. Gluteus medius
E. Tensor fascia latae

Group B: In the blank to the left of each origin description, write the **LETTER** of the muscle with that origin.

1. _____ Ischial tuberosity only
2. _____ Anterior body of pubis
3. _____ Inferior ramus and body of pubis
4. _____ **Anterior head:** Anterior inferior iliac spine
   Posterior head: upper margin of acetabulum
5. _____ Inferior ramus of pubis, ramus of ischium, and ischial tuberosity
6. _____ Anterior superior iliac spine and iliac notch
7. _____ **Long head:** ischial tuberosity
   Short head: lateral lip of linea aspera and supracondylar line
8. _____ Intertrochanteric line, greater trochanter, and gluteal tuberosity

A. Sartorius
B. Rectus femoris
C. Vastus lateralis
D. Biceps femoris
E. Semitendinosus
F. Gracilis
G. Adductor longus
H. Adductor magnus

Group C: In the blank to the left of each insertion description, write the **LETTER** of the muscle with that insertion.

1. _____ Tuberosity of navicular, dorsal surfaces of cuboid, cuneiforms, and base of metatarsals
2. _____ Plantar surface of bases of distal phalanges of toes 2–5
3. _____ Dorsal surface of base of fifth metatarsal bone
4. _____ Proximal posterior surface of tibia
5. _____ Calcaneus through Achilles tendon
6. _____ Lateral side of medial cuneiform and base of first metatarsal
7. _____ Plantar surface of base of distal phalanx of big toe
8. _____ Medial plantar surface of first cuneiform and base of first metatarsal

A. Gastrocnemius
B. Popliteus
C. Flexor digitorum longus
D. Flexor hallucis longus
E. Tibialis posterior
F. Tibialis anterior
G. Fibularis longus
H. Fibularis tertius

III. SENTENCE COMPLETION

Circle the term in parentheses that correctly completes each statement.

1. All of the following muscles flex the thigh at the hip joint *except* the (**iliacus, psoas, gluteus maximus, tensor fascia latae**).
2. Many muscles work together to rotate the thigh laterally at the hip. All of the following muscles *except* the (**obturator internus, piriformis, gluteus minimus, quadratus femoris**) rotate the hip laterally.
3. The tendon of the (**gluteus maximus, sartorius, semitendinosus, tensor fascia latae**) muscle forms the iliotibial band.
4. The strap-like muscle that crosses both the hip and knee joints and is used when sitting in a yoga position is the (**sartorius, gracilis, tensor fascia latae, rectus femoris**).
5. All of the following muscles *except* the (**biceps femoris, rectus femoris, vastus lateralis, vastus intermedius**) extend the leg at the knee.
6. The hamstring muscles consist of all of the following *except* the (**biceps femoris, rectus femoris, semitendinosus, semimembranosus**).
7. Chondromalacia patellae, or runner's knee, can be caused by imbalance of the (**gastrocnemius and soleus, hamstring muscles, quadriceps muscles**).
8. The muscle that is the prime mover in both hip extension and knee flexion is the (**sartorius, biceps femoris, rectus femoris, gracilis**).
9. All of the following muscles *except* the (**pectineus, sartorius, gracilis, adductor brevis**) adduct the thigh.
10. The group of muscles that are important in horseback riding or other activities that require pressing the thighs together is the (**hamstrings, quadriceps, triceps surae, adductors**).

11. All of the following muscles medially rotate the hip at the thigh *except* the (**gluteus maximus, gluteus medius, tensor fascia latae, gracilis**).

12. The muscle that plantar flexes the foot and also assists in flexion of the knee is the (**tibialis anterior, soleus, gastrocnemius, plantaris**).

13. All of the following muscles are antagonistic to the gastrocnemius *except* the (**soleus, tibialis anterior, extensor digitorum longus, fibularis tertius**).

14. The tendon of the (**plantaris, popliteus, soleus, tensor fascia latae**) muscle can be removed and used to reconstruct the tendons of the hand.

15. The deepest muscle in the back of the knee is the (**plantaris, fibularis tertius, tibialis posterior, popliteus**).

16. The muscle that flexes the distal phalanges of toes 2 through 5 is the (**flexor hallucis longus, flexor digitorum longus, flexor digitorum brevis, fibularis brevis**).

17. Paralysis of the (**tibialis posterior, fibularis longus, gastrocnemius, tibialis anterior**) causes "foot drop," and irritation of it causes "shin splints."

18. All of the following muscles *except* the (**fibularis longus, tibialis anterior, fibularis brevis, extensor digitorum longus**) are evertors of the foot.

19. The muscles of the sole of the foot can be divided into four layers. The most superficial layer consists of all of the following *except* the (**lumbricales, flexor digitorum brevis, abductor digiti minimi, abductor hallucis**).

20. The muscle that adducts the big toe is the (**adduct digitorum, adductor hallucis, lumbricales, adductor digiti minimi**).

IV. MULTIPLE CHOICE

In the blank to the left of each question, write the **LETTER** of the correct answer.

1. _____ The popliteus muscle _________ the leg at the knee.
   A. Adducts
   B. Abducts
   C. Extends
   D. Medially rotates

2. _____ Overstretching of the gracilis and adductors muscles of the thigh commonly results from
   A. Soccer
   B. Tennis
   C. Horseback riding
   D. Running

3. _____ Which of the following major muscles is involved in crossing the leg at the knee?
   A. Gastrocnemius
   B. Rectus femoris
   C. Semimembranosus
   D. Sartorius

4. _____ What muscle both plantar flexes and everts the foot?
   A. Tibialis anterior
   B. Gastrocnemius
   C. Plantaris
   D. Fibularis longus

5. _____ Sciatic nerve damage diminishes ability to do which of the following movements?
   A. Flex the hip
   B. Flex the knee
   C. Adduct the hip
   D. Abduct the hip

6. _____ Which muscles are involved when massaging the thigh in the supine position?
   A. Hamstrings
   B. Quadriceps
   C. Gluteals
   D. Triceps surae

7. _____ The iliopsoas flexes the hip because of its insertion on the
   A. Greater trochanter
   B. Iliac crest
   C. Lesser trochanter
   D. Ischial tuberosity

8. _____ Which of the following muscles does *not* attach to the os coxae?
   A. Gracilis
   B. Sartorius
   C. Vastus medialis
   D. Semimembranosus

9. _____ Standing on tiptoes requires what movement at the ankle joint?
   A. Dorsiflexion
   B. Eversion
   C. Plantar flexion
   D. Inversion

10. _____ Which of the following muscles does *not* help flex the hip?
    A. Rectus femoris
    B. Sartorius
    C. Iliopsoas
    D. Semimembranosus

11. _____ The gluteus medius inserts on the ________
    A. Greater trochanter
    B. Medial epicondyle
    C. Gluteal tuberosity
    D. Linea aspera

12. _____ Which muscle of the foot is the deepest?
    A. Plantar interossei
    B. Adductor hallucis
    C. Quadratus plantae
    D. Abductor hallucis

13. _____ Two common sites in the leg for intramuscular injection are the
    A. Gluteus maximus and sartorius
    B. Gluteus medius and rectus femoris
    C. Gluteus maximus and vastus medialis
    D. Gluteus medius and semimembranosus

14. _____ The quadriceps muscle that originates on the os coxa is the
    A. Vastus intermedius
    B. Vastus lateralis
    C. Vastus medialis
    D. Rectus femoris

15. _____ The most lateral hamstring muscle is the
    A. Biceps femoris
    B. Rectus femoris
    C. Semimembranosus
    D. Semitendinosus

16. _____ All of the following adductor muscles *except* the __________ insert at least in part on the linea aspera.
    A. Adductor magnus
    B. Gracilis
    C. Pectineus
    D. Adductor longus

17. _____ The tendon of this muscle can be palpated on the medial malleolus during active inversion of foot.
    A. Tibialis anterior
    B. Tibialis posterior
    C. Flexor hallucis longus
    D. Soleus

18. _____ Antagonists to the tibialis anterior include all of the following *except*
    A. Gastrocnemius
    B. Soleus
    C. Extensor digitorum longus
    D. Fibularis longus

19. _____ The triceps surae is composed of the _________ and _________ muscles.
    A. Triceps brachii and anconeus
    B. Gastrocnemius and soleus
    C. Popliteus and plantaris
    D. Vastus lateralis and vastus medialis

20. _____ All of the small hip muscles *except* the _________ laterally rotate the thigh.
    A. Obturator externus
    B. Piriformis
    C. Gluteus minimus
    D. Gemellus superior

V. FILL IN THE BLANK

Write the correct word(s) in the blank spaces to complete each statement.

1. The muscle that flexes the little toe is the ________________ .

2. One of the muscles that helps maintain the transverse arch of the foot is the ________________ .

3. One of the muscles that flexes the terminal phalanges of toes 2 through 5 is the ________________ .

4. Dorsiflexion and eversion of the foot is caused by the ________________ and ________________ muscles.

5. The muscle that abducts the big toe is the ________________ .

6. The prime mover in dorsiflexion of the foot is the ________________ .

7. The muscle that flexes the big toe is the ________________ , and the muscle that extends the big toe is the ________________ .

8. The triceps surae muscles insert on the ________________ .

9. The muscle that is the prime mover in plantar flexion of the foot and that assists in the flexion at the knee is the ________________ .

10. The femoral artery and vein pass through a hiatal separation in the ________________ muscle.

11. The most proximal of the muscles that adduct the thigh is the ________________ muscle.

12. A "pulled groin" injury involves injury to several muscles in the ________________ group.

13. The three hamstring muscles are the ________________ , ________________ , and ________________ .

14. The quadriceps muscle that originates in part from the linea aspera is the ________________ .

15. A deep muscle bruise within the fascia of the muscle is called a ________________ .

16. The four muscles in the quadriceps group are the ________________ , ________________ , ________________ , and ________________ .

17. A "high pointer" injury may involve both a bruise and an avulsion on the iliac crest caused by one of the attached muscles. One of the muscles attached to the iliac crest is the ________________ .

18. The two gluteal muscles that medially rotate the thigh at the hip are the ______________ and ______________ .

19. The gemellus muscles, quadratus femoris, and obturator muscles ______________ rotate the hip at the thigh.

20. The sciatic nerve may become entrapped as it passes through the ______________ muscle.

VI. SHORT ANSWER

Answer each of the questions in the space provided.

1. The term "groin strain" often involves several muscles. Which muscles are most likely to be involved in a groin strain? What type of activity might cause groin strain?

2. What is chondromalacia patella? What causes this injury?

3. What are the symptoms of a ruptured Achilles tendon? What muscles would be impaired? What specific foot actions would not be able to be performed?

4. What is "foot drop"? What muscle is not functional?

5. The muscles of the lower legs are typically divided into three compartments for ease of study. What are these three compartments? What is the primary action produced by simultaneous contraction of the muscles in each compartment?

VII. CASE STUDIES

A. While kicking a soccer ball, a 20-year-old man experienced a sharp pain and later found a prominent lump in the upper medial portion of his left thigh. A soft mass could be palpated. Below the mass a shallow space in the musculature lateral to the gracilis and medial to the sartorius was found. Computed tomographic (CT) examination indicated that the mass was muscular.

1. A member of what muscle group probably made up the mass?

2. Which muscle in the group was most likely affected?

3. What would cause the muscle to bulge?

4. What are some other possible causes of a mass in this region?

B. A new runner experienced pain in her anterior lower left leg. As she continued to work out, the pain increased. Her leg was red, swollen, and tender. Dorsiflexion of her foot was painful, and sensation was lost between the first two toes.

1. What is the probable diagnosis for her condition?

2. Which muscle is primarily involved?

3. Which nerve is being compromised?

4. Paralysis of the nerve in question 3 would cause what condition?

Functional Muscle Groups

6

At the end of this chapter, you should be able to

1. Identify the muscles in the elevators and depressors, protractors and retractors, and upward and downward rotators of the scapula.
2. Identify the muscles in the medial and lateral rotators, flexors and extensors, and adductors and abductors of the humerus.
3. Identify the muscles in the flexors and extensors of the elbow.
4. Identify the muscles in the pronators and supinators of the forearm.
5. Identify the muscles in the flexors and extensors and adductors and abductors of the wrist.
6. Identify the muscles in the adductors and abductors, flexors and extensors, and opposition of the thumb and digits.
7. Identify the muscles in the medial and lateral rotators, flexors and extensors, adductors and abductors of the hip.
8. Identify the muscles in the medial and lateral rotators, flexors, and extensors of the knee.
9. Identify the muscles in the dorsiflexors and plantar flexors of the ankle.
10. Identify the muscles in the invertors and evertors of the foot.

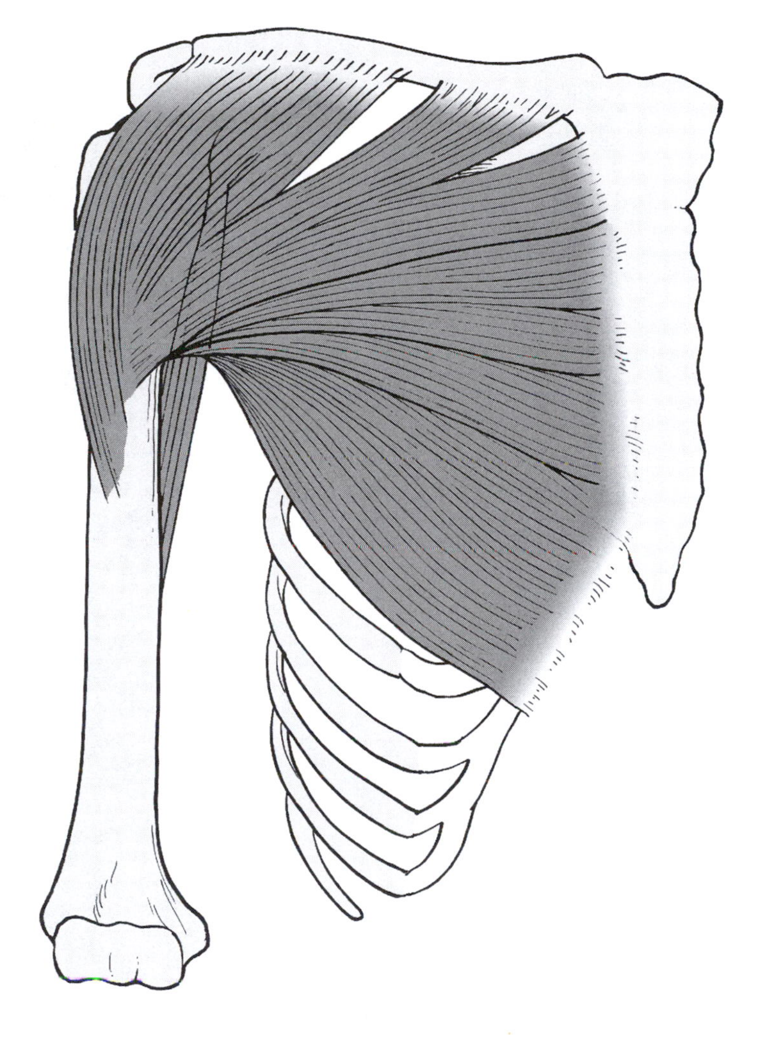

Elevators of Scapula

(ell•a•**vay**•tors) (**skap**•yoo•lah)

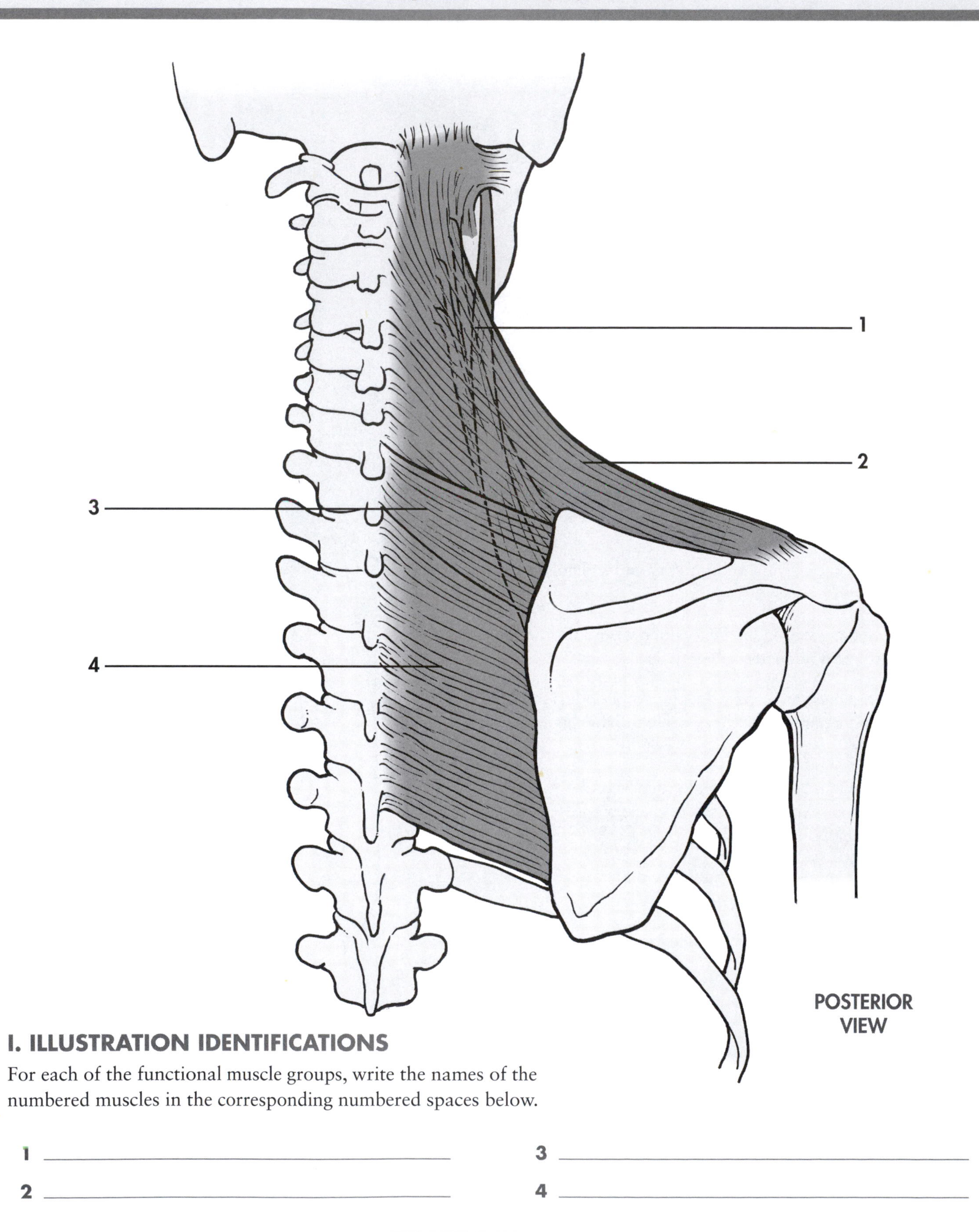

I. ILLUSTRATION IDENTIFICATIONS

For each of the functional muscle groups, write the names of the numbered muscles in the corresponding numbered spaces below.

1 ______________________ 3 ______________________

2 ______________________ 4 ______________________

Depressors of Scapula

(de•**press**•ors) (**skap**•yoo•lah)

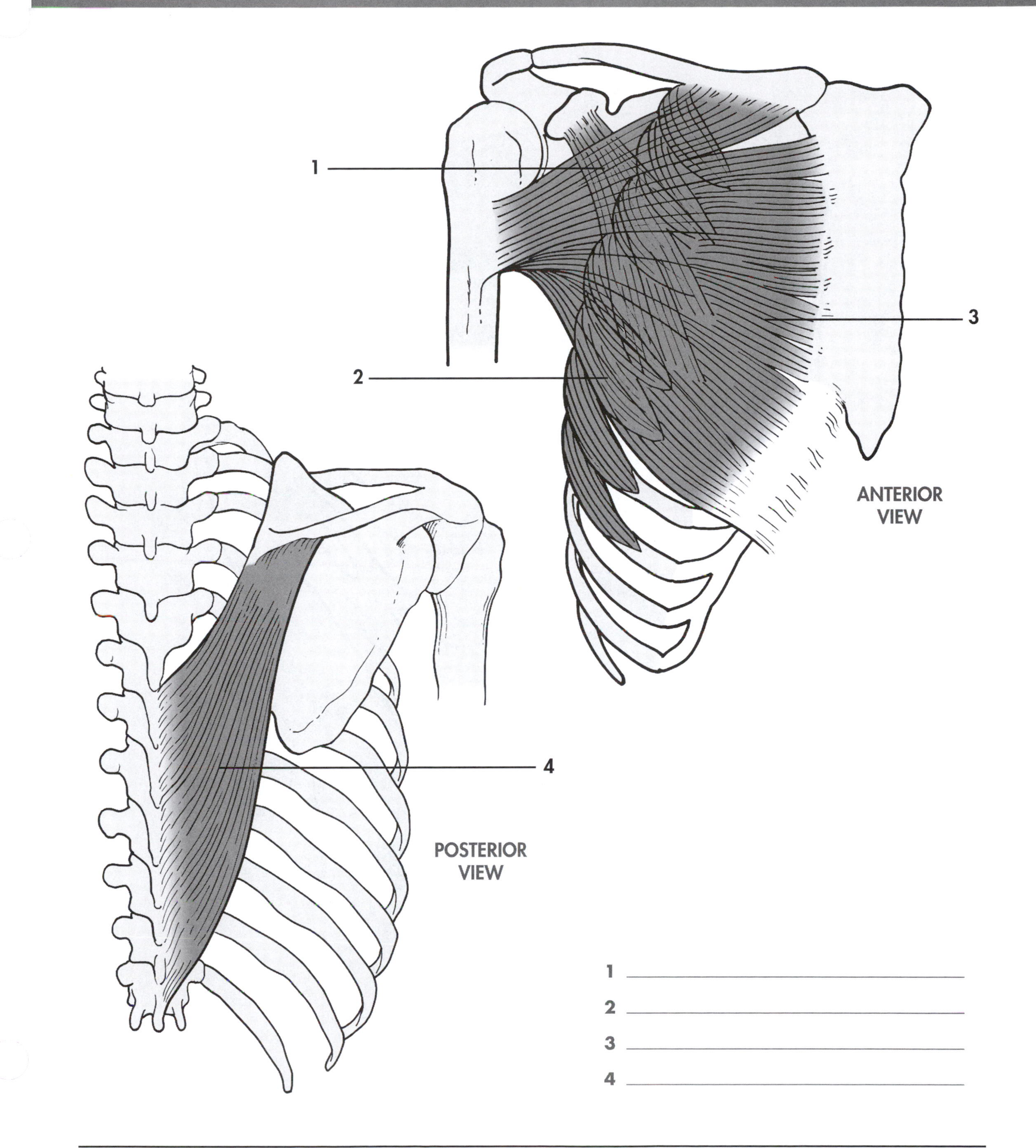

Protractors of Scapula

(pro•**track**•tors) (**skap**•yoo•lah)

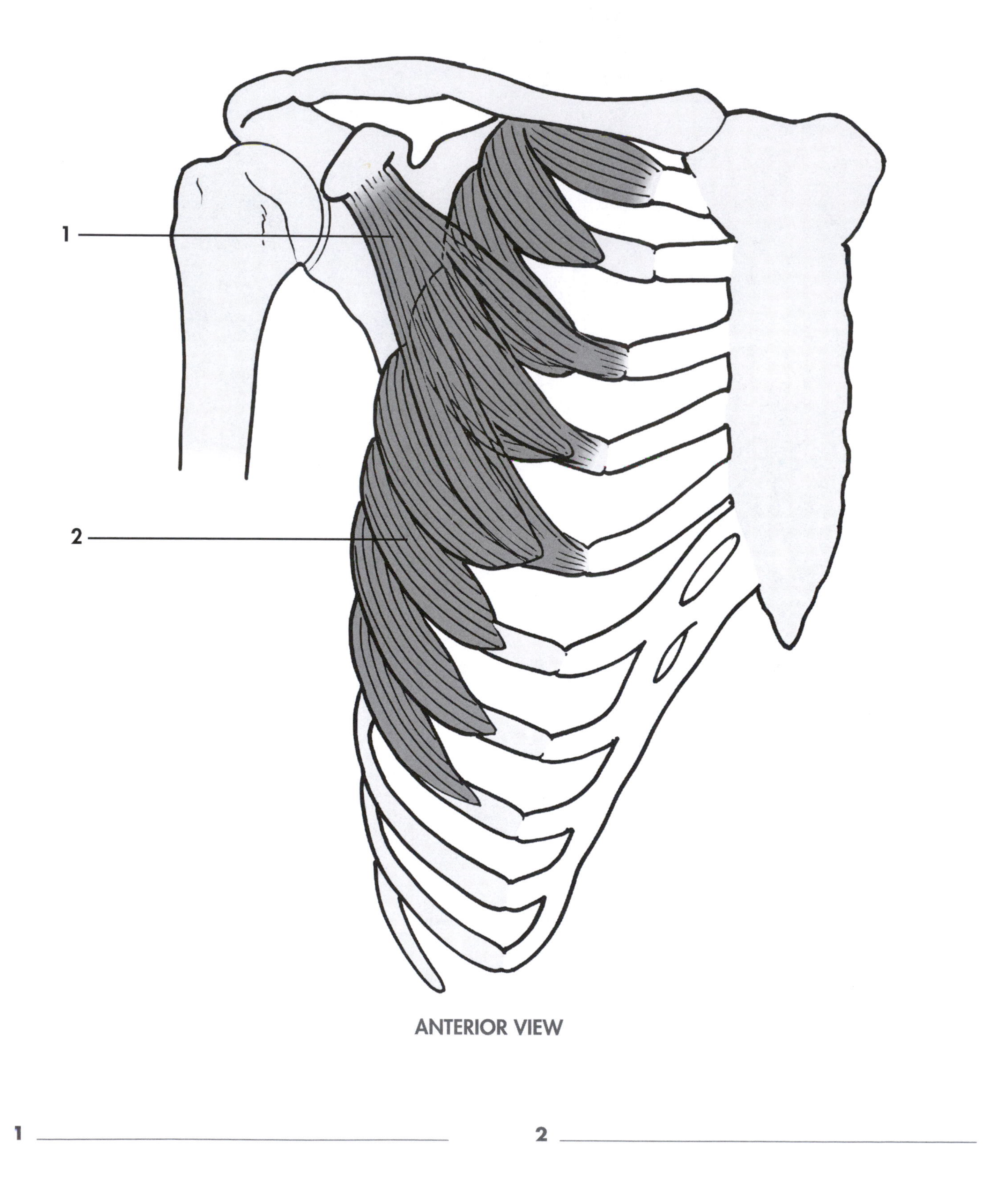

ANTERIOR VIEW

1 ______________________ 2 ______________________

Retractors of Scapula

(re•**track**•tors) (**skap**•yoo•lah)

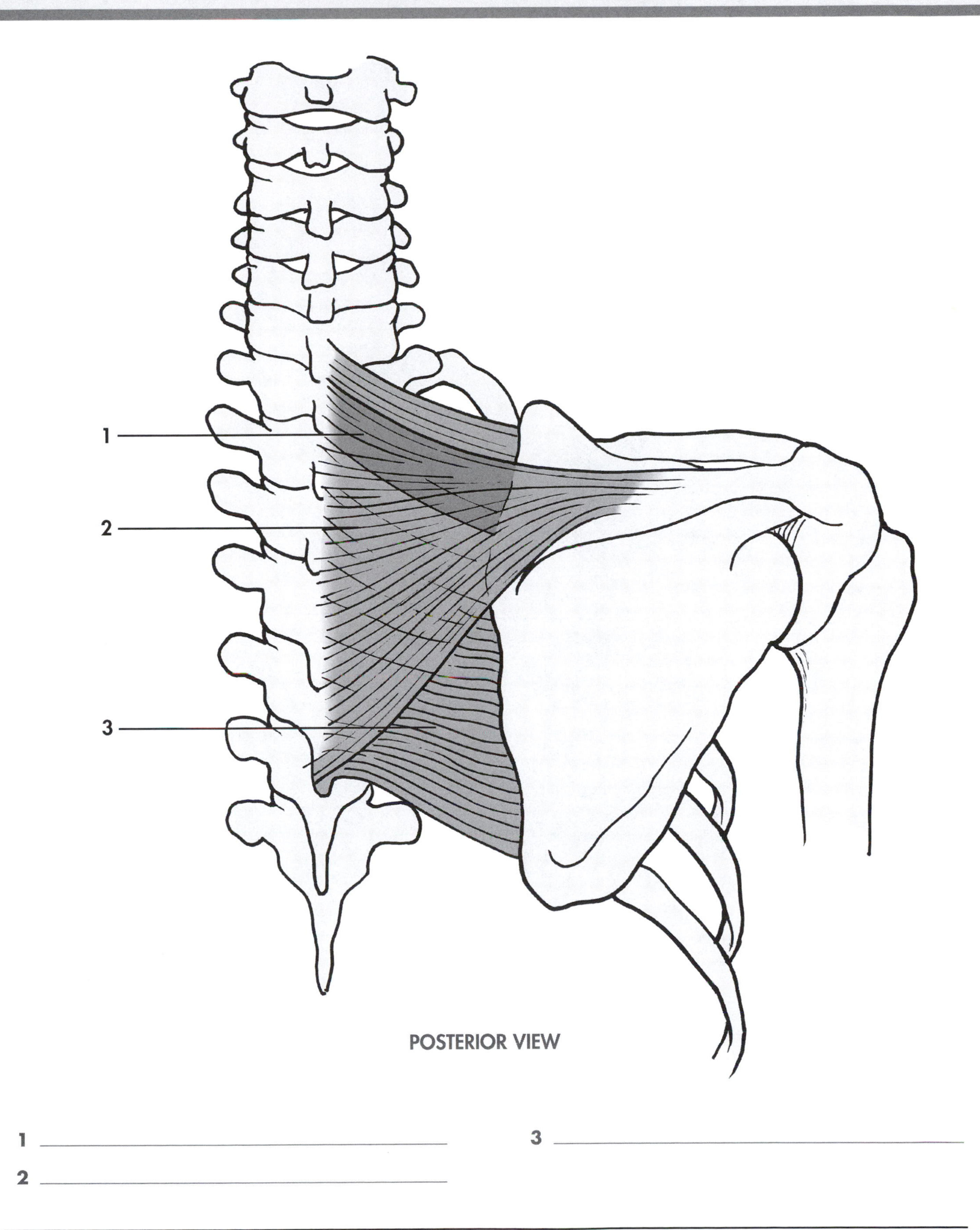

POSTERIOR VIEW

1 ______________________

2 ______________________

3 ______________________

Upward Rotators of Scapula

(ro•**tay**•tors) (**skap**•yoo•lah)

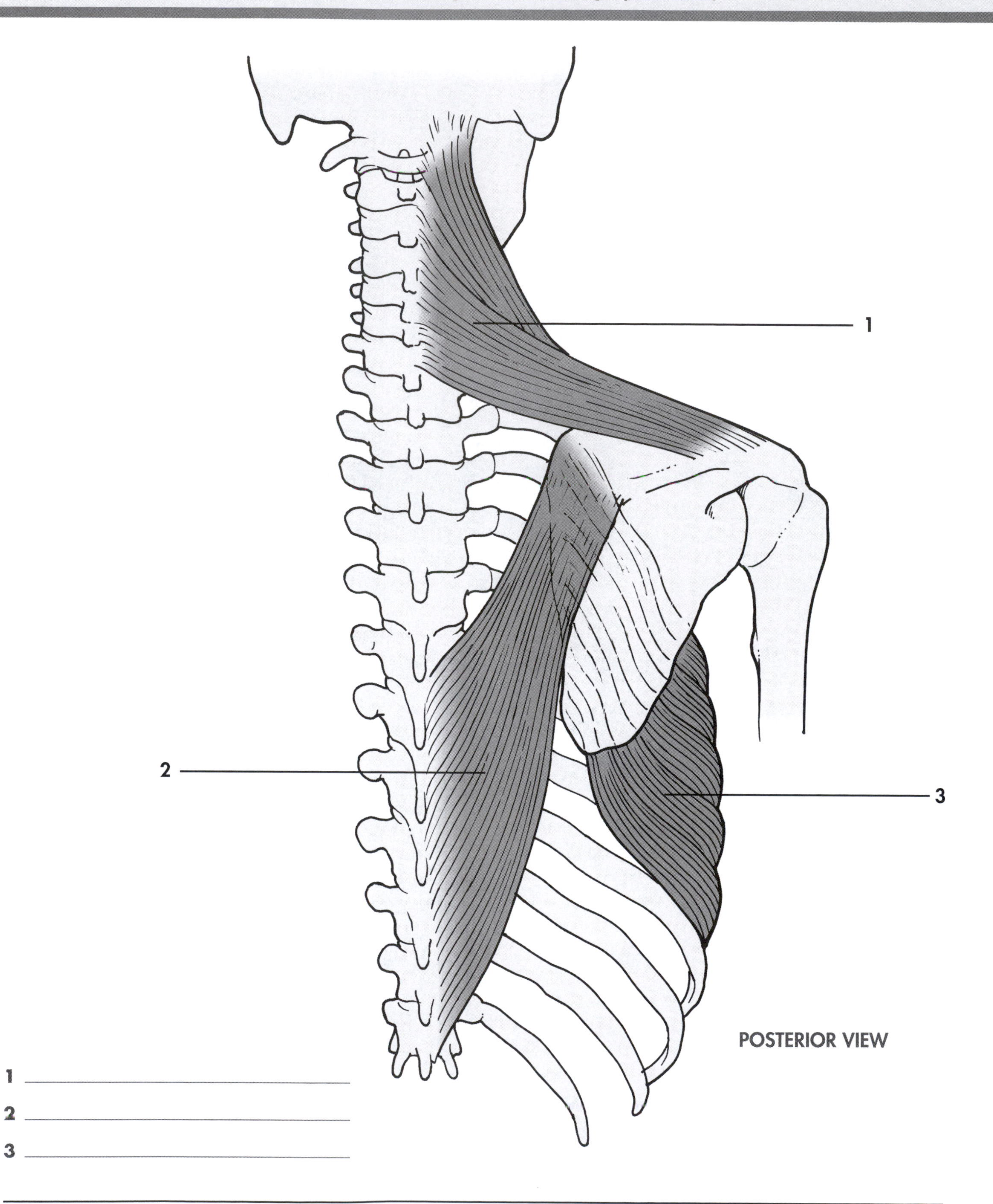

1 ______________________________

2 ______________________________

3 ______________________________

Downward Rotators of Scapula

(ro•**tay**•tors) (**skap**•yoo•lah)

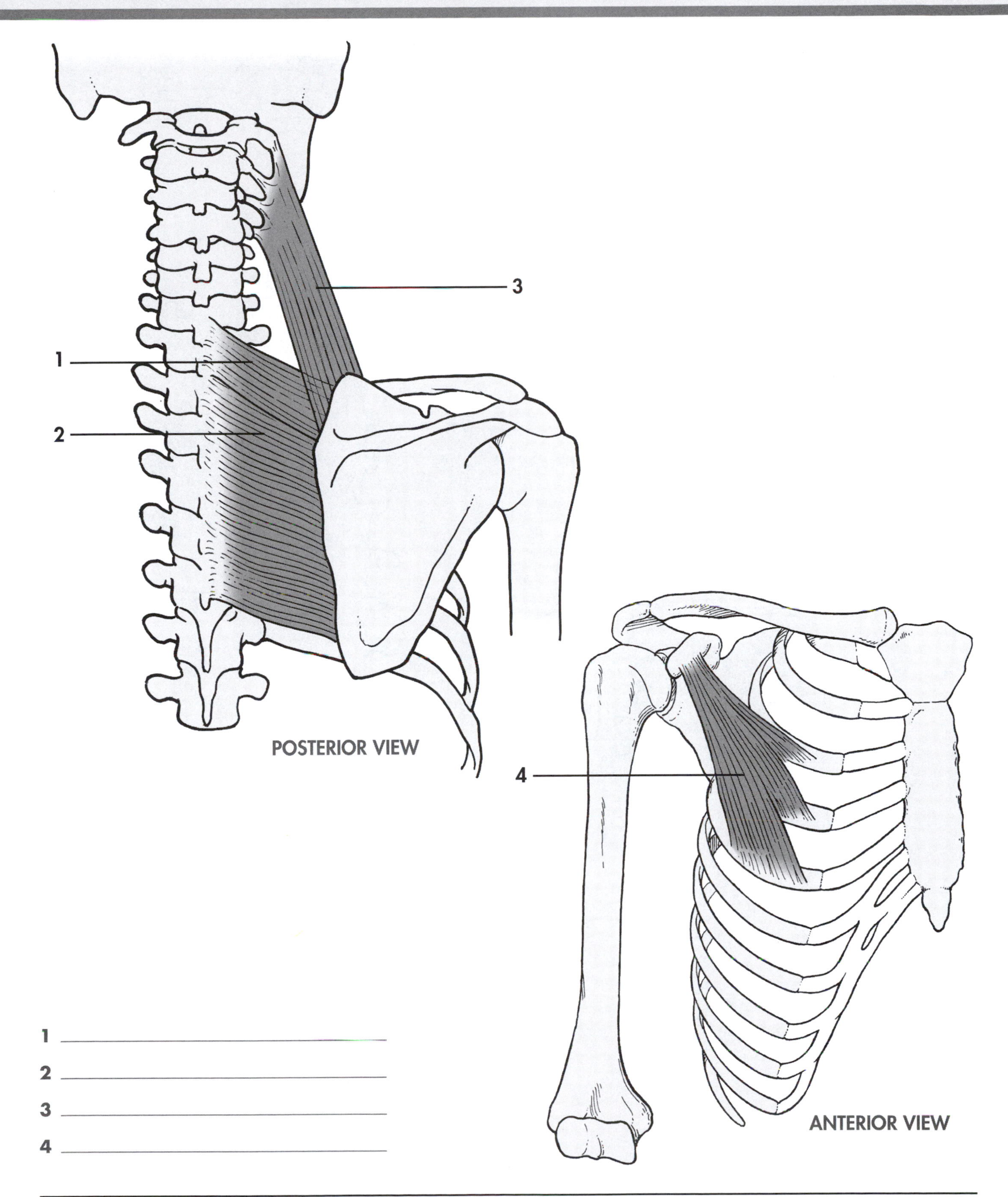

1 ______________________

2 ______________________

3 ______________________

4 ______________________

Medial Rotators of Humerus

(ro•**tay**•tors) (**hyoo**•mir•us)

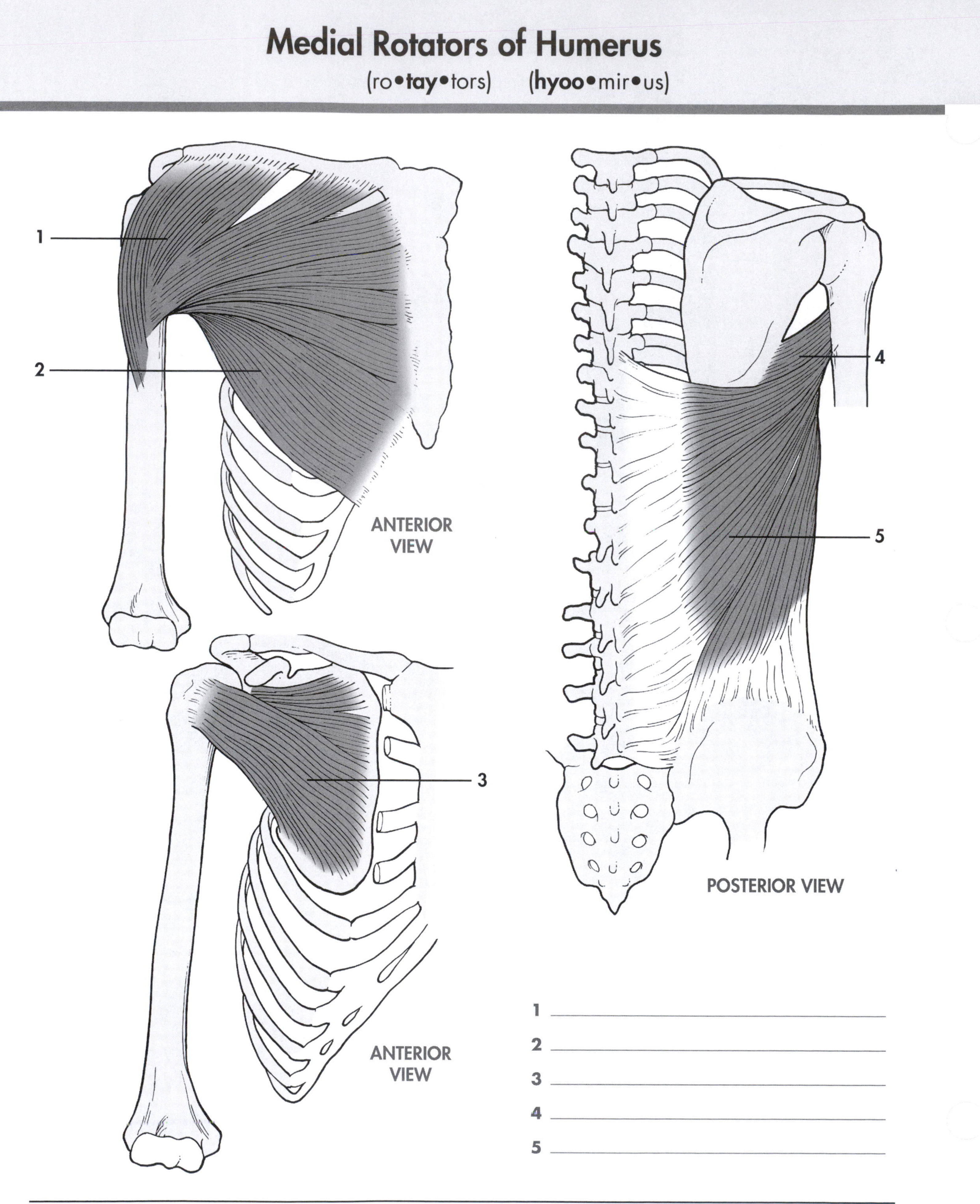

1 ______________________

2 ______________________

3 ______________________

4 ______________________

5 ______________________

Lateral Rotators of Humerus

(ro•**tay**•tors) (**hyoo**•mir•us)

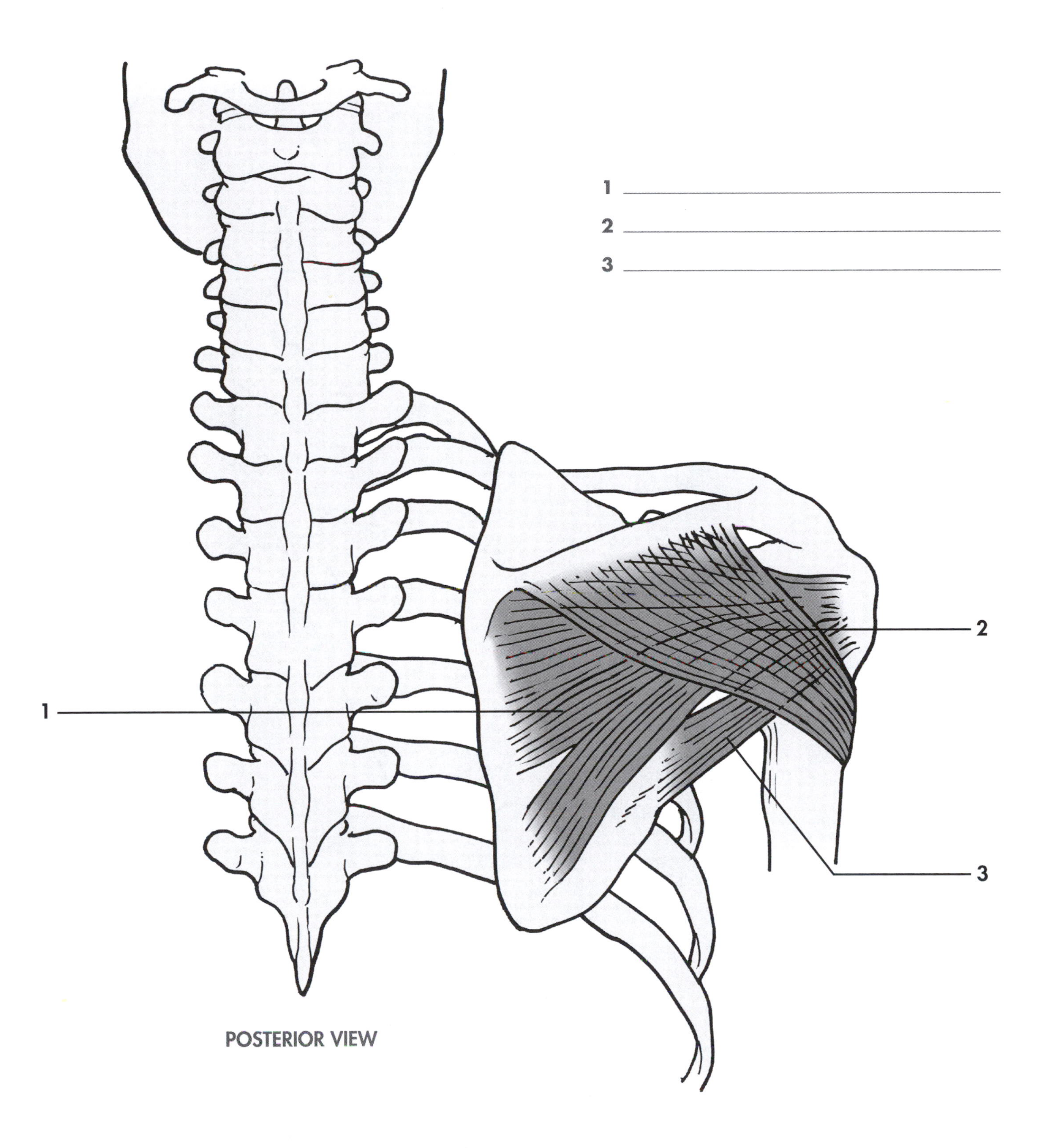

Flexors of Humerus

(flek•sors) (hyoo•mir•us)

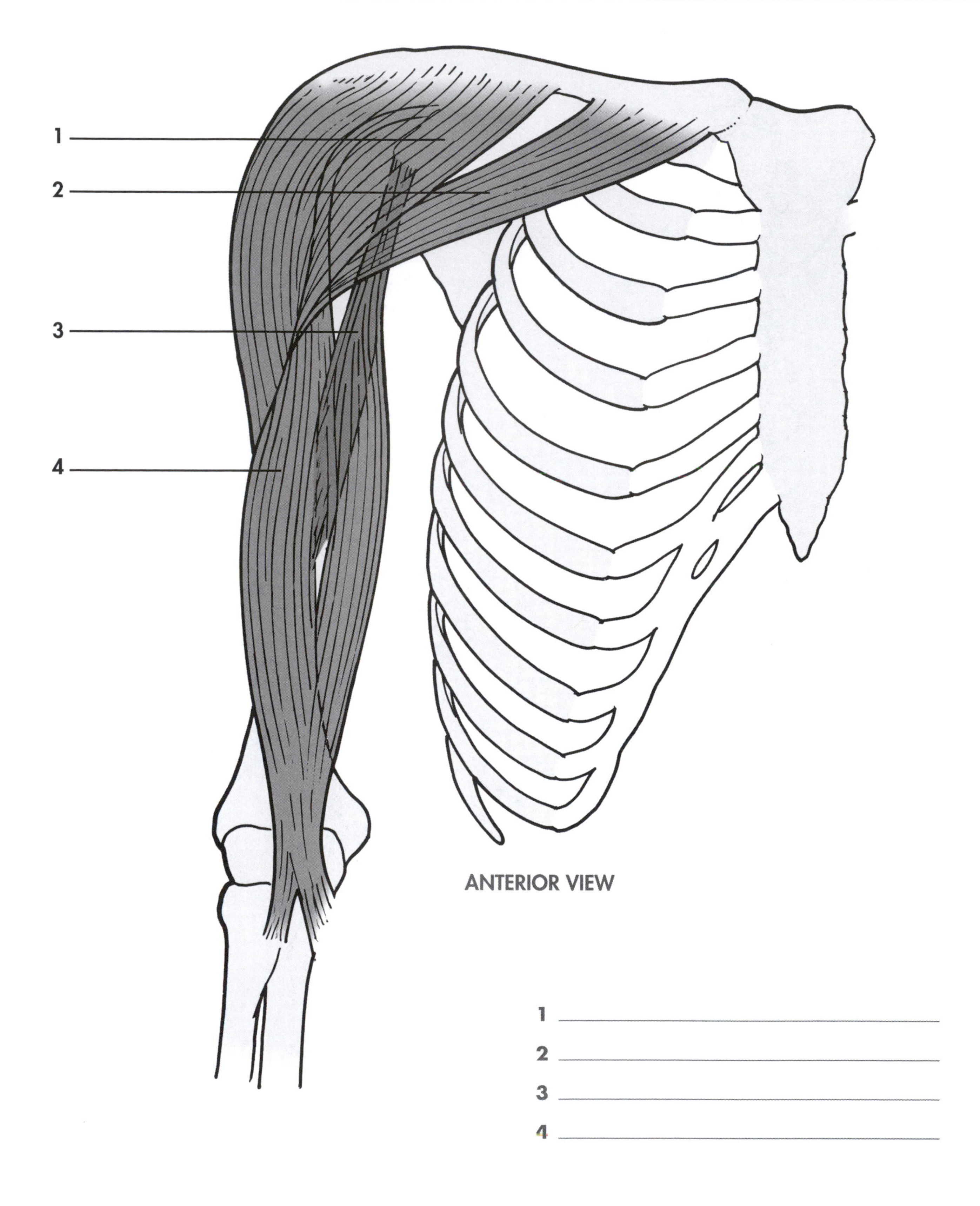

1 ____________________

2 ____________________

3 ____________________

4 ____________________

Extensors of Humerus

(ex•**sten**•sors) (**hyoo**•mir•us)

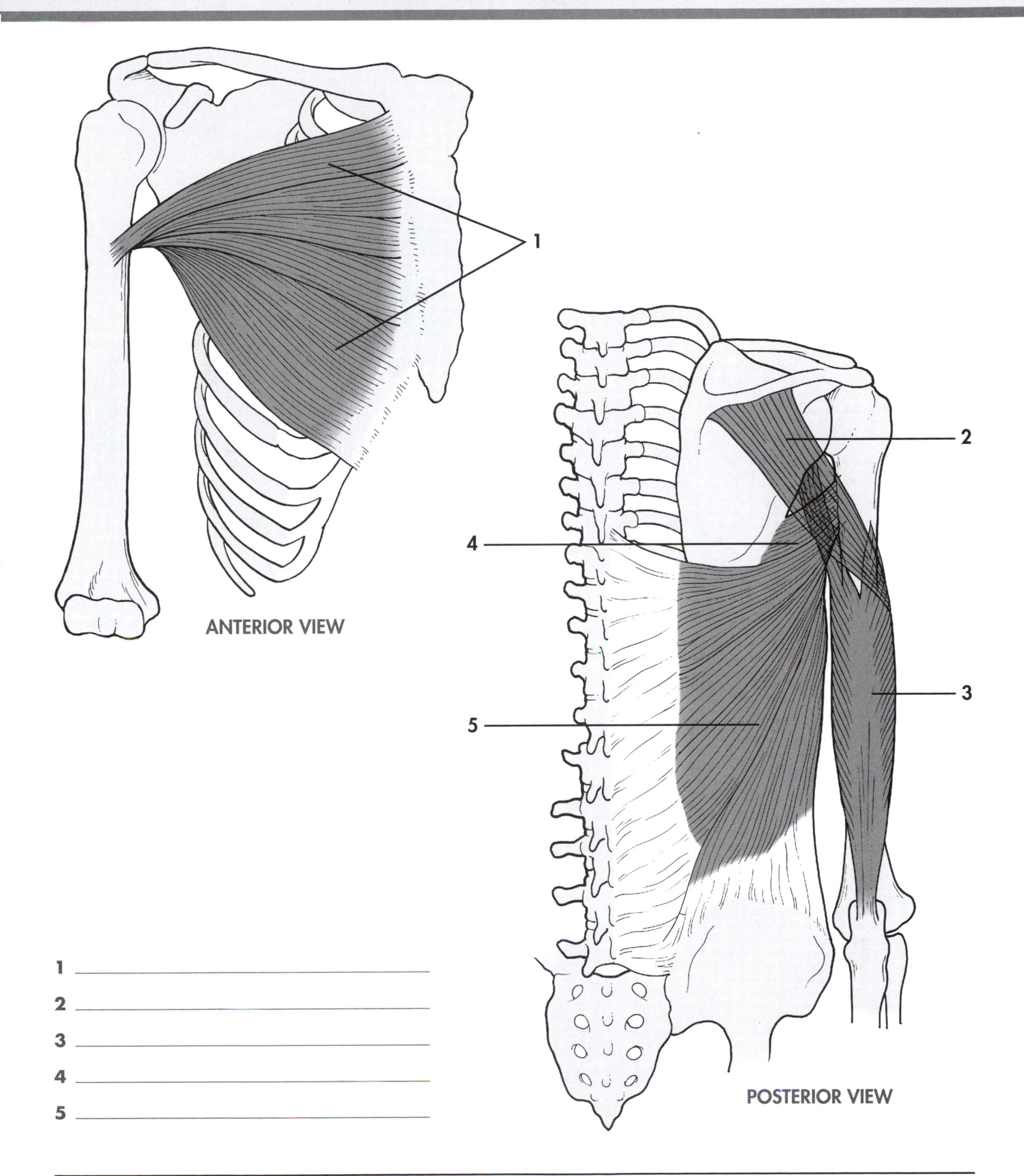

1 ______________________

2 ______________________

3 ______________________

4 ______________________

5 ______________________

Abductors of Humerus

(**ab**•duck•tors) (**hyoo**•mir•us)

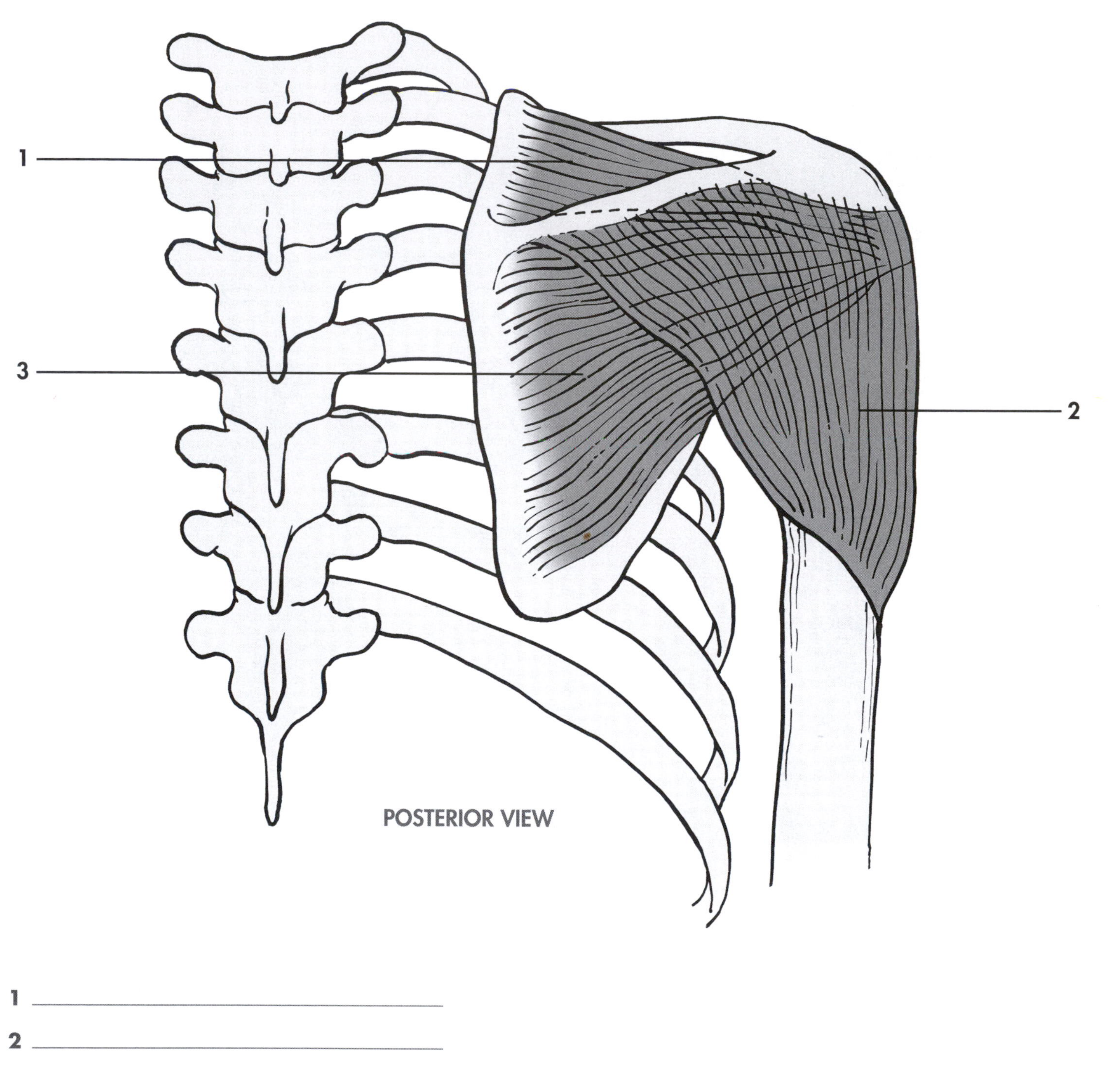

1 ______________________________

2 ______________________________

3 ______________________________

Adductors of Humerus

(**ad**•duck•tors) (**hyoo**•mir•us)

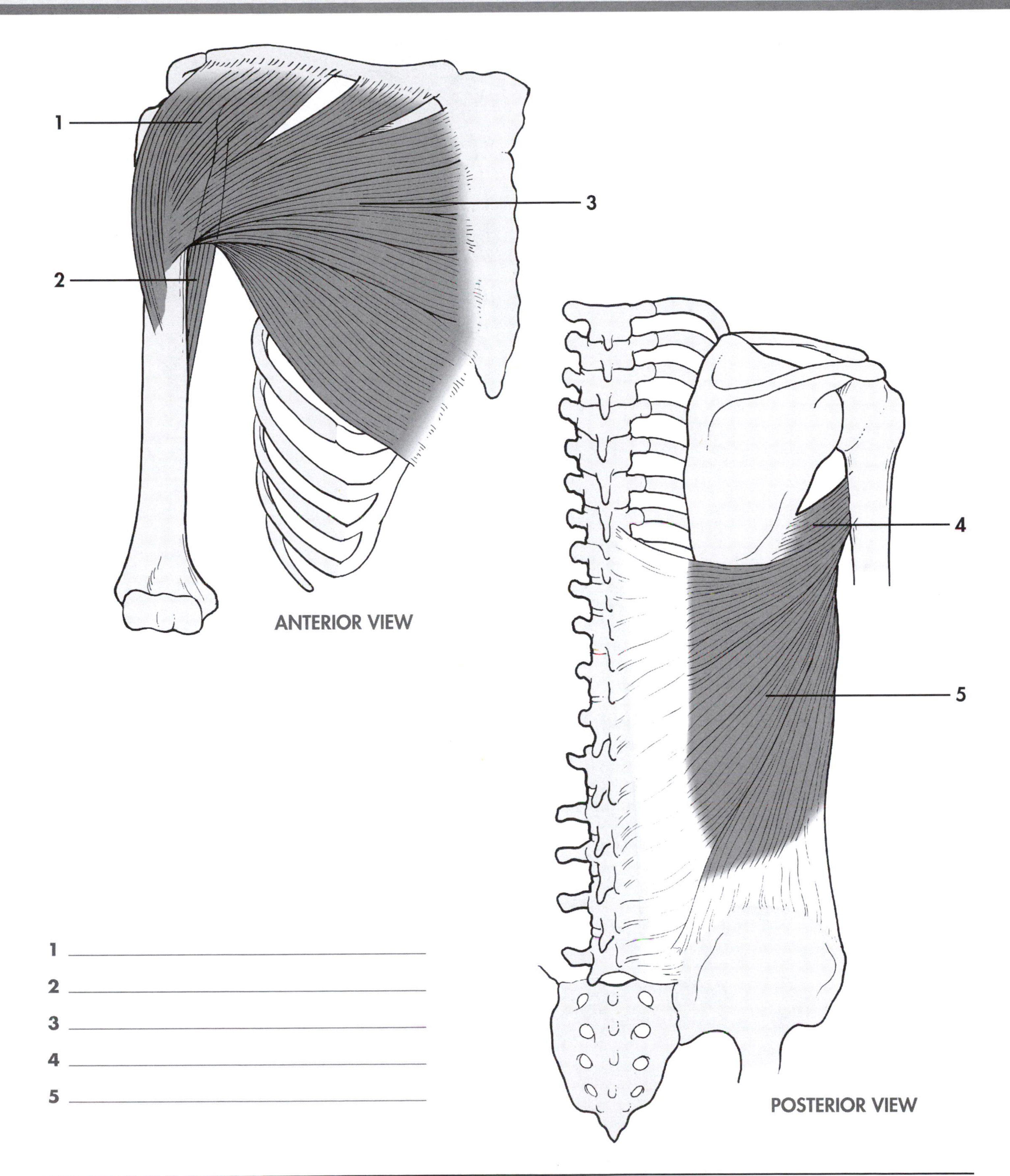

Flexors of Elbow

(flek•sors)

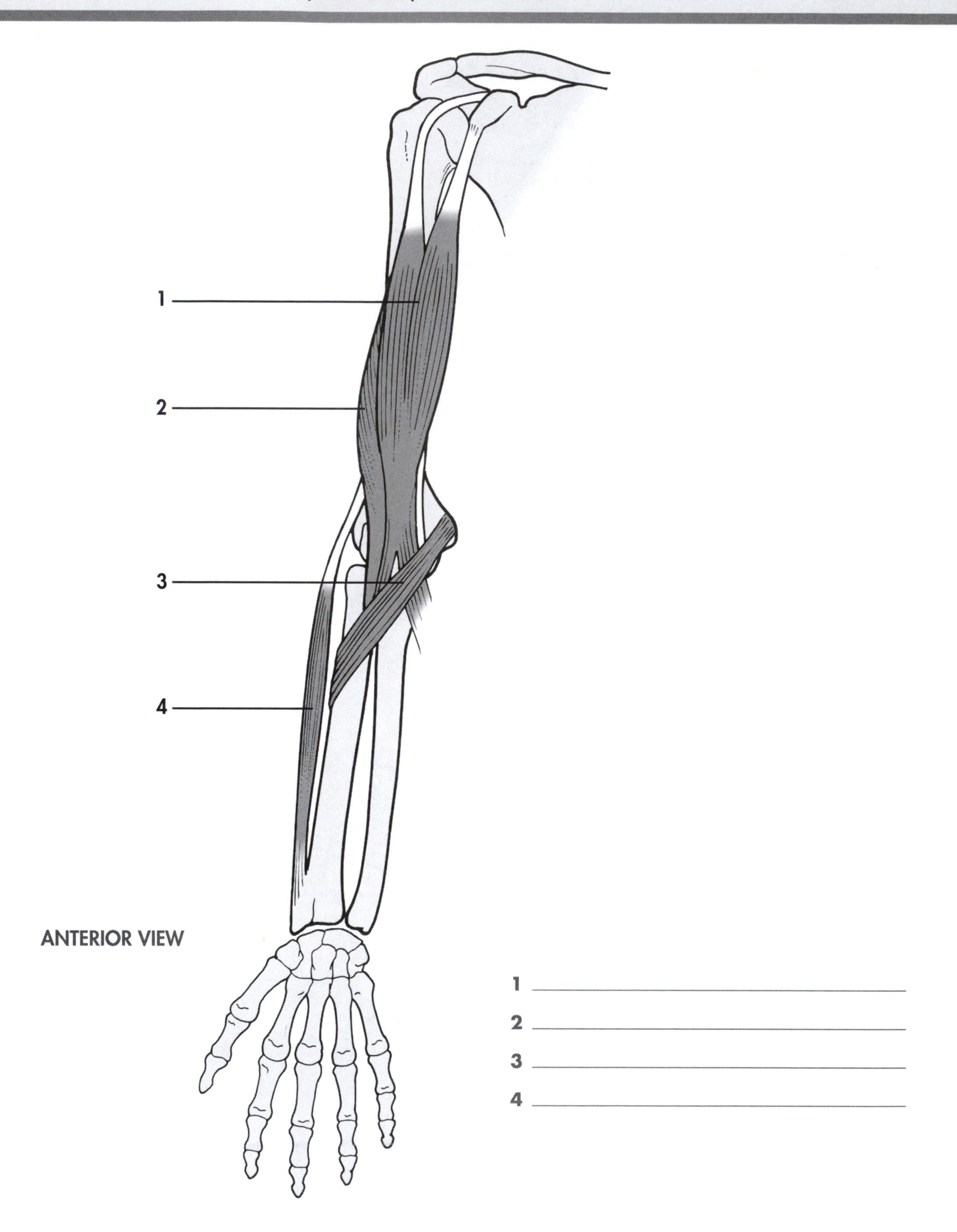

1 ______________________

2 ______________________

3 ______________________

4 ______________________

Extensors of Elbow

(ex•**sten**•sors)

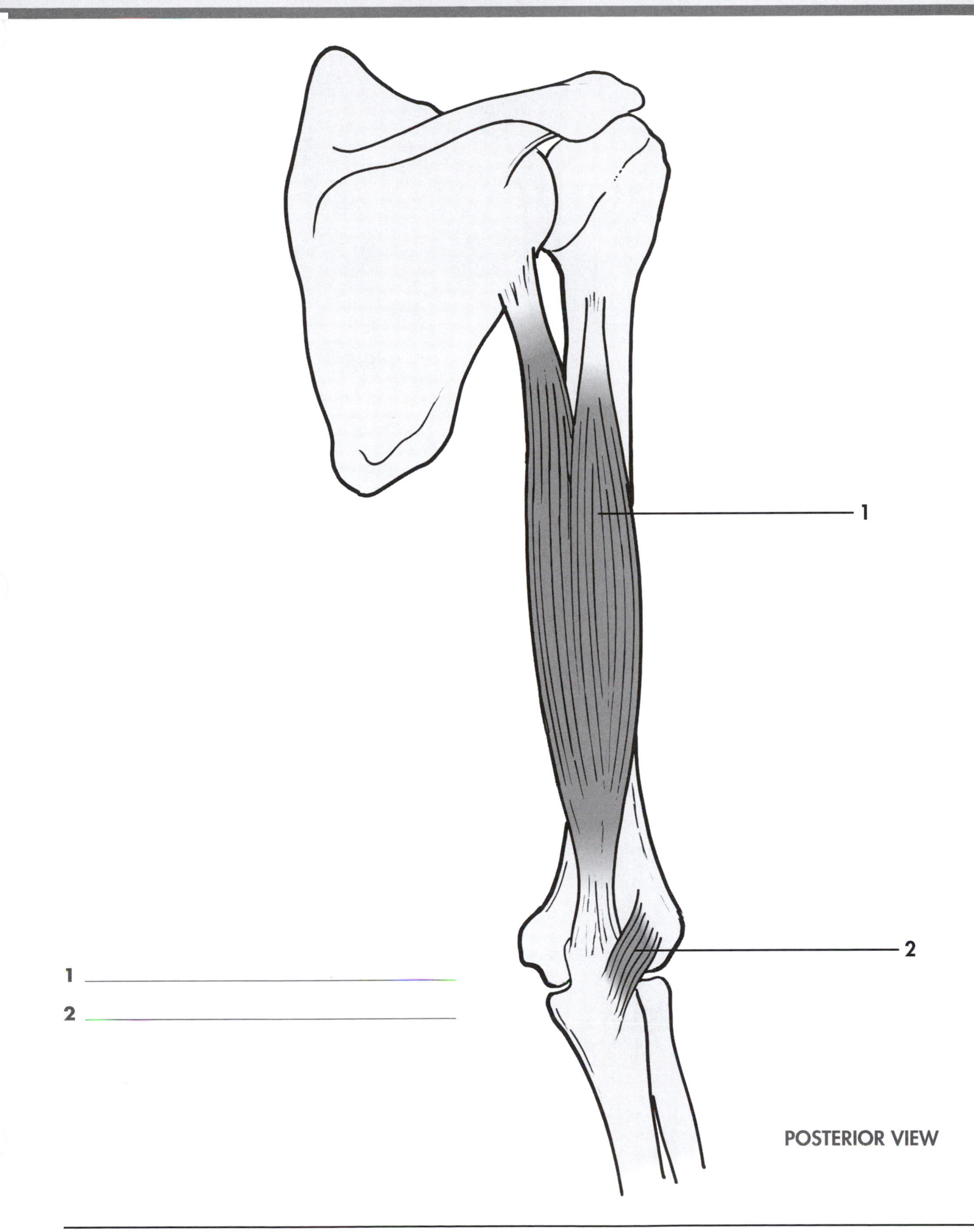

1 ______________________

2 ______________________

Supinators of Forearm

(soo•pin•**nay**•tors)

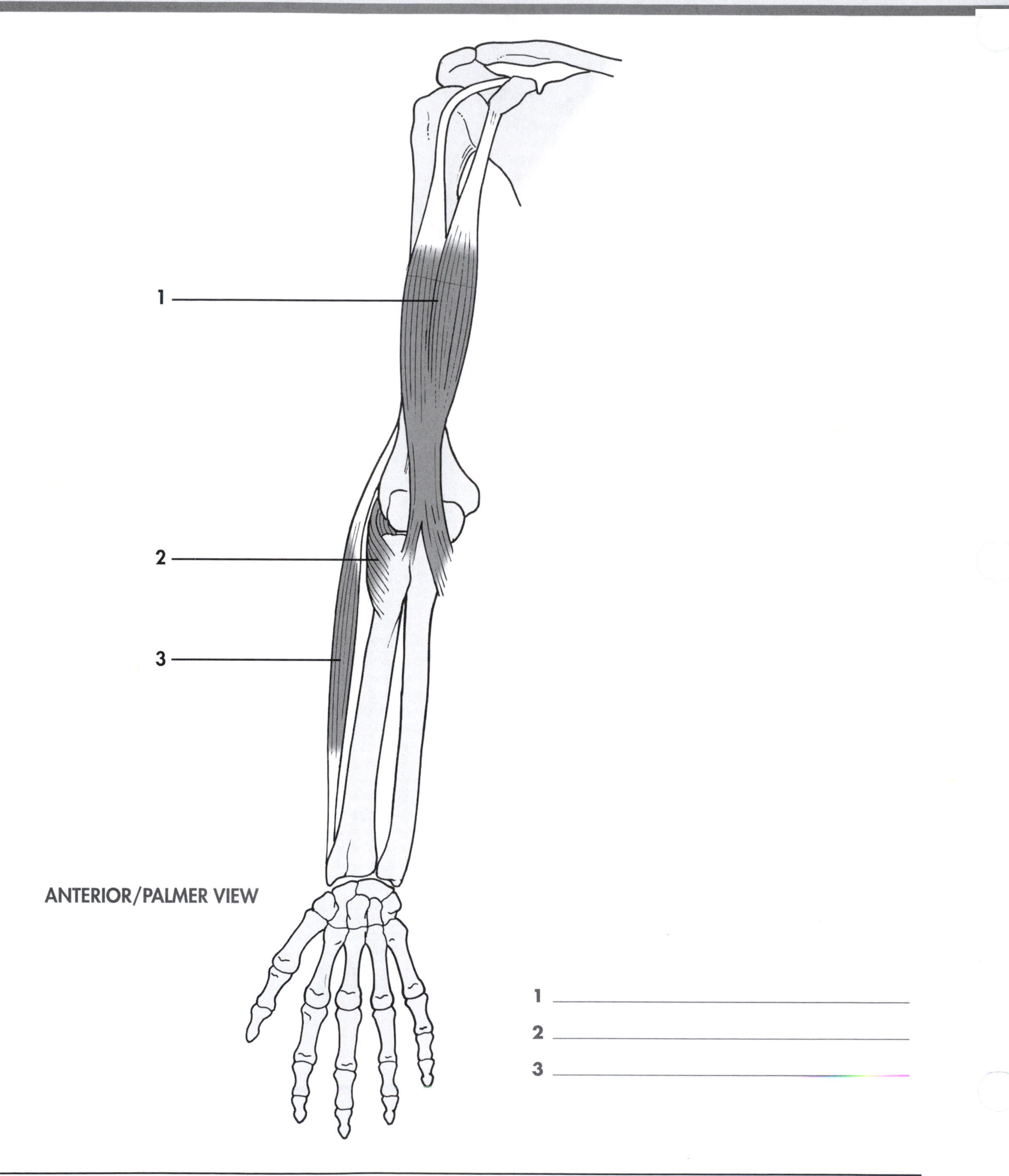

Pronators of Forearm

(pro•**nay**•tors)

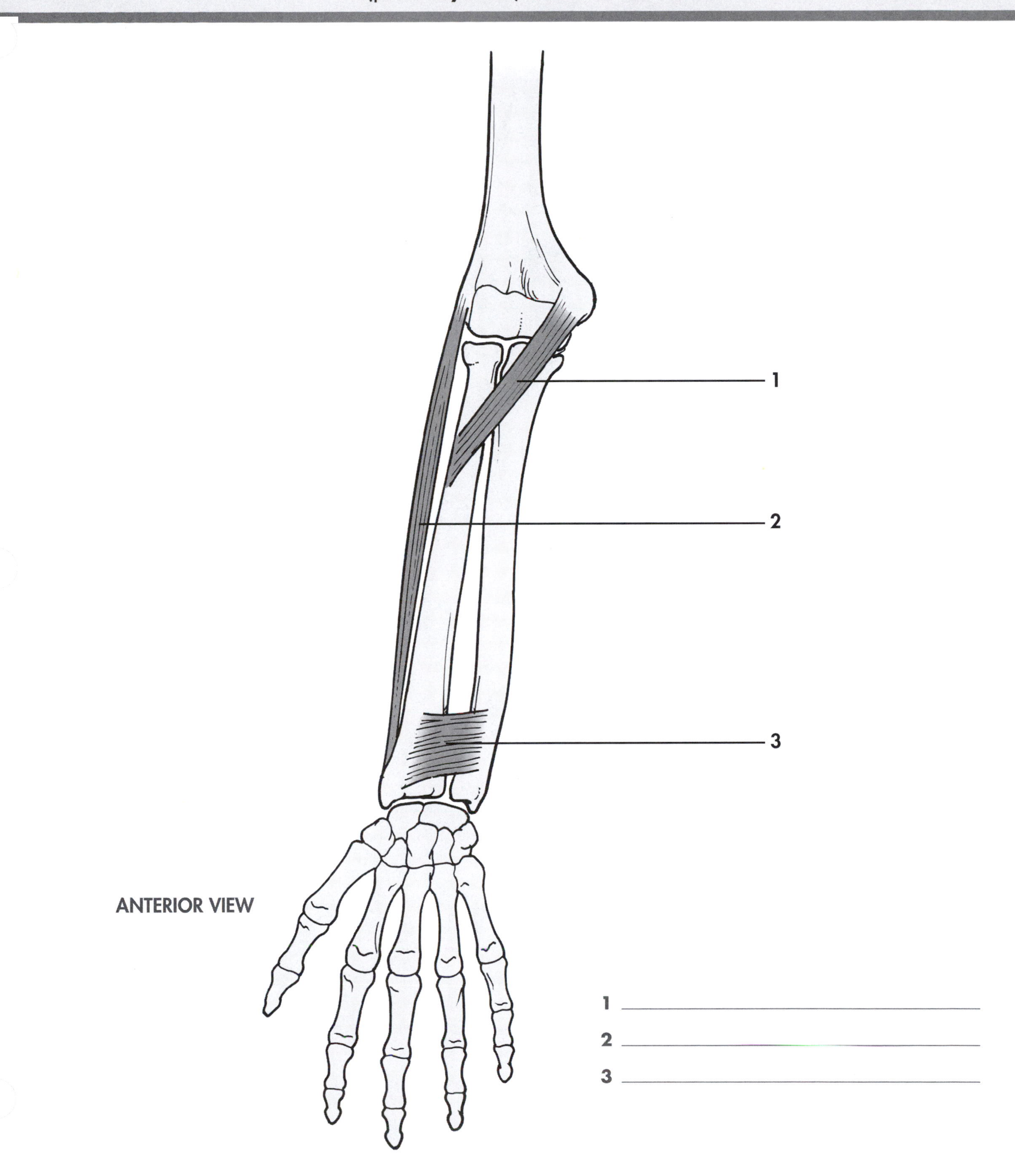

Flexors of Wrist

(**flek**•sors)

1 ______________________________

2 ______________________________

3 ______________________________

4 ______________________________

5 ______________________________

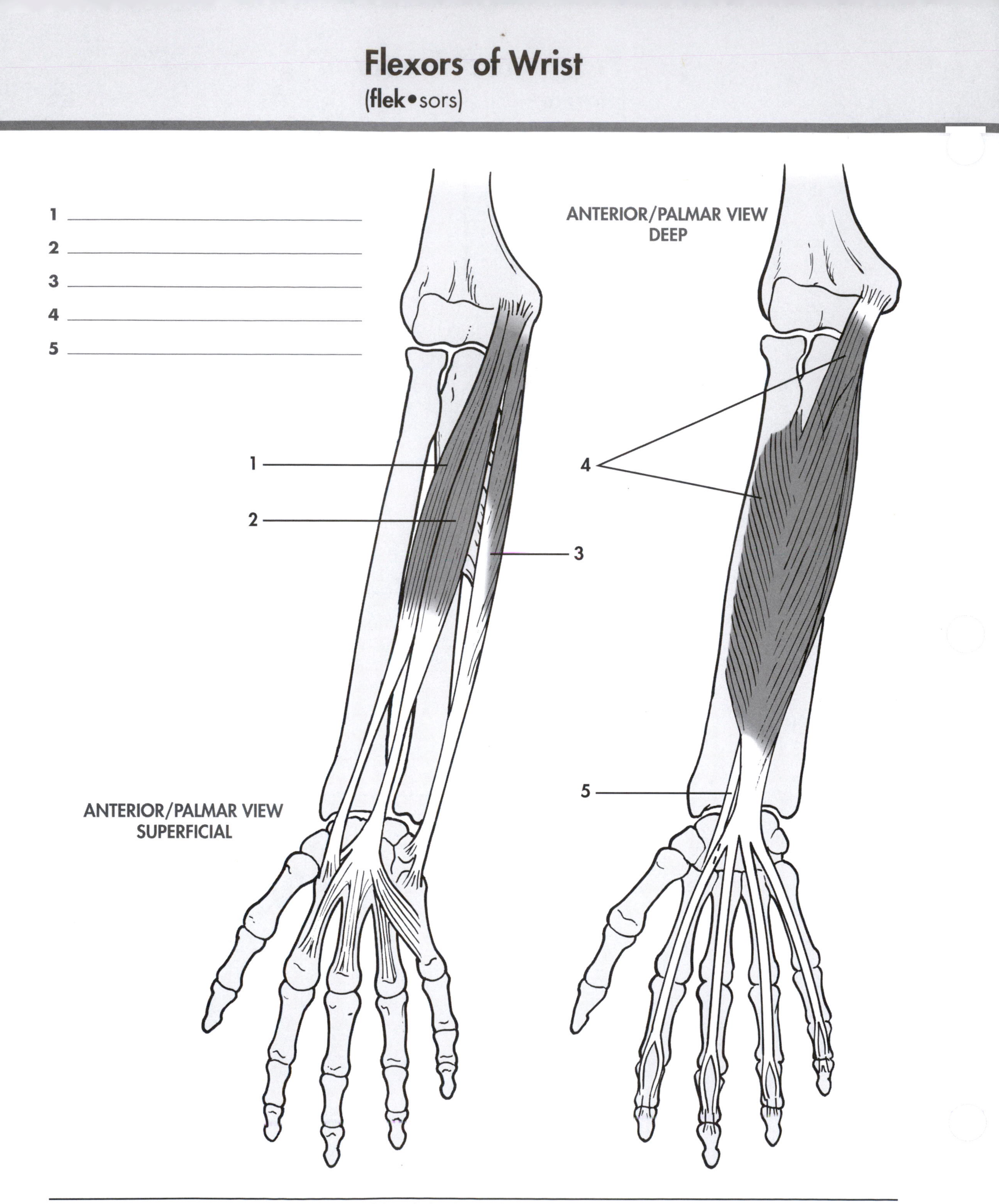

Extensors of Wrist

(ex•**sten**•sors)

1 ______________________________

2 ______________________________

3 ______________________________

4 ______________________________

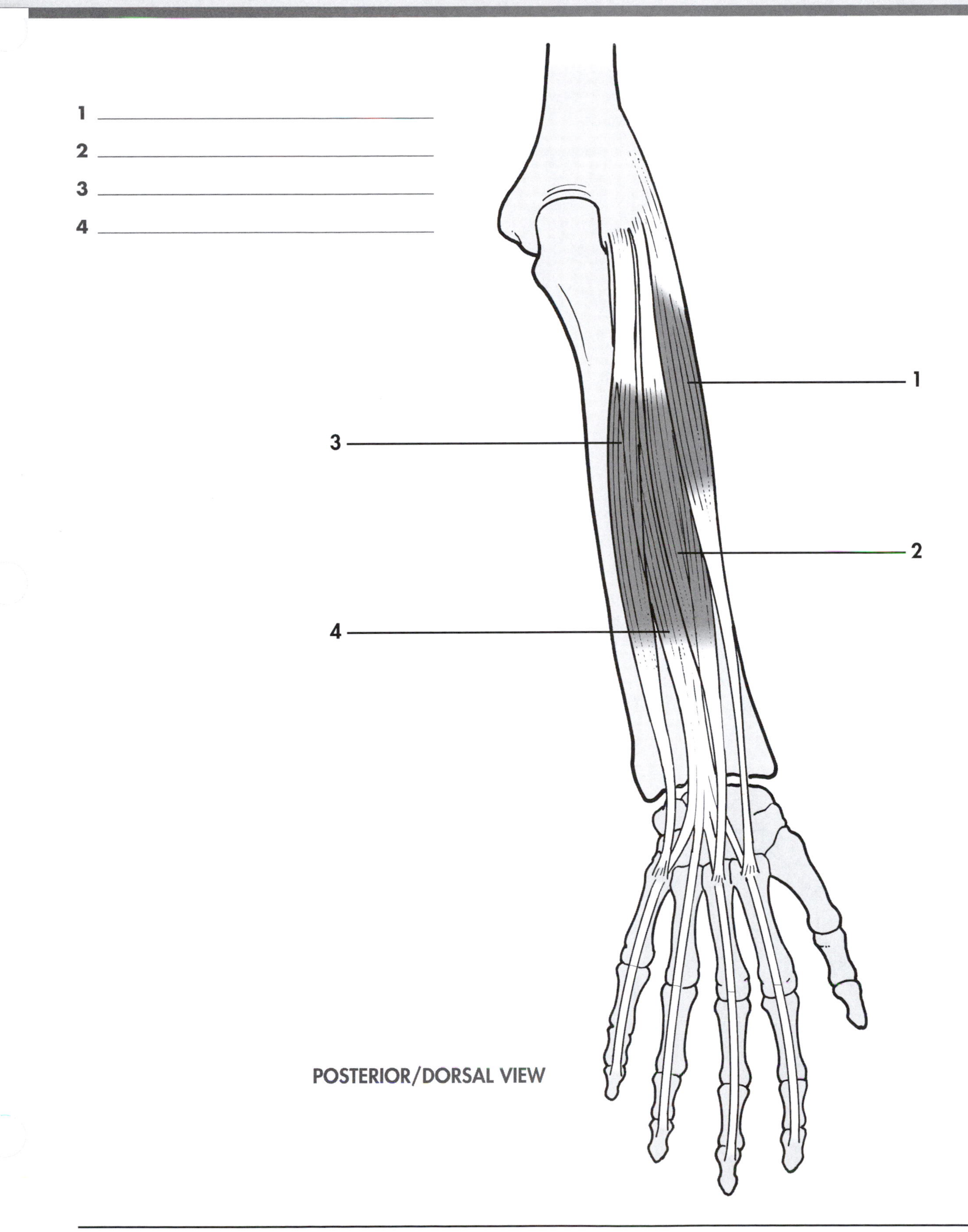

POSTERIOR/DORSAL VIEW

Adducors of Wrist

(**ad**•duck•tors)

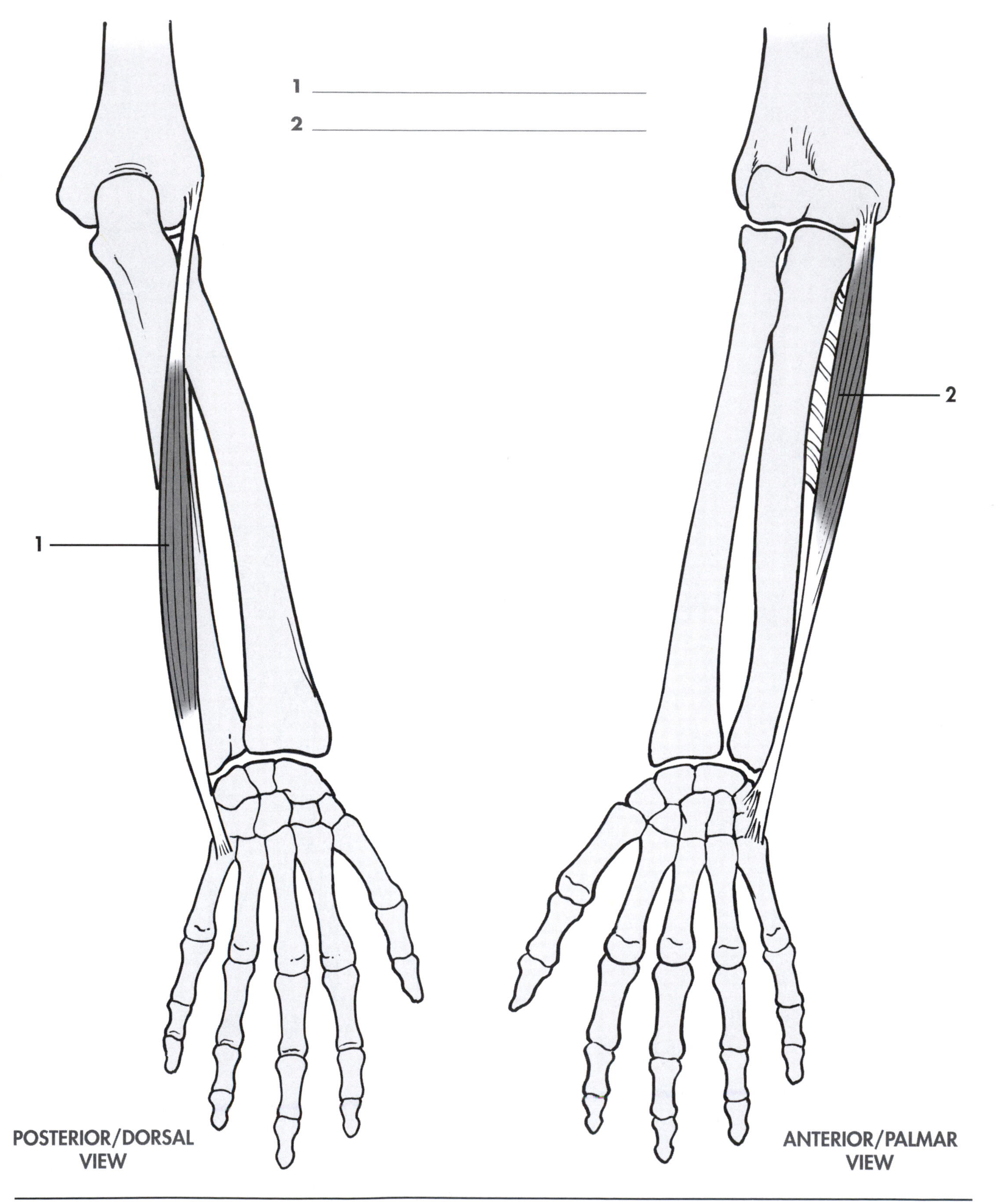

Abductors of Wrist

(ab•duck•tors)

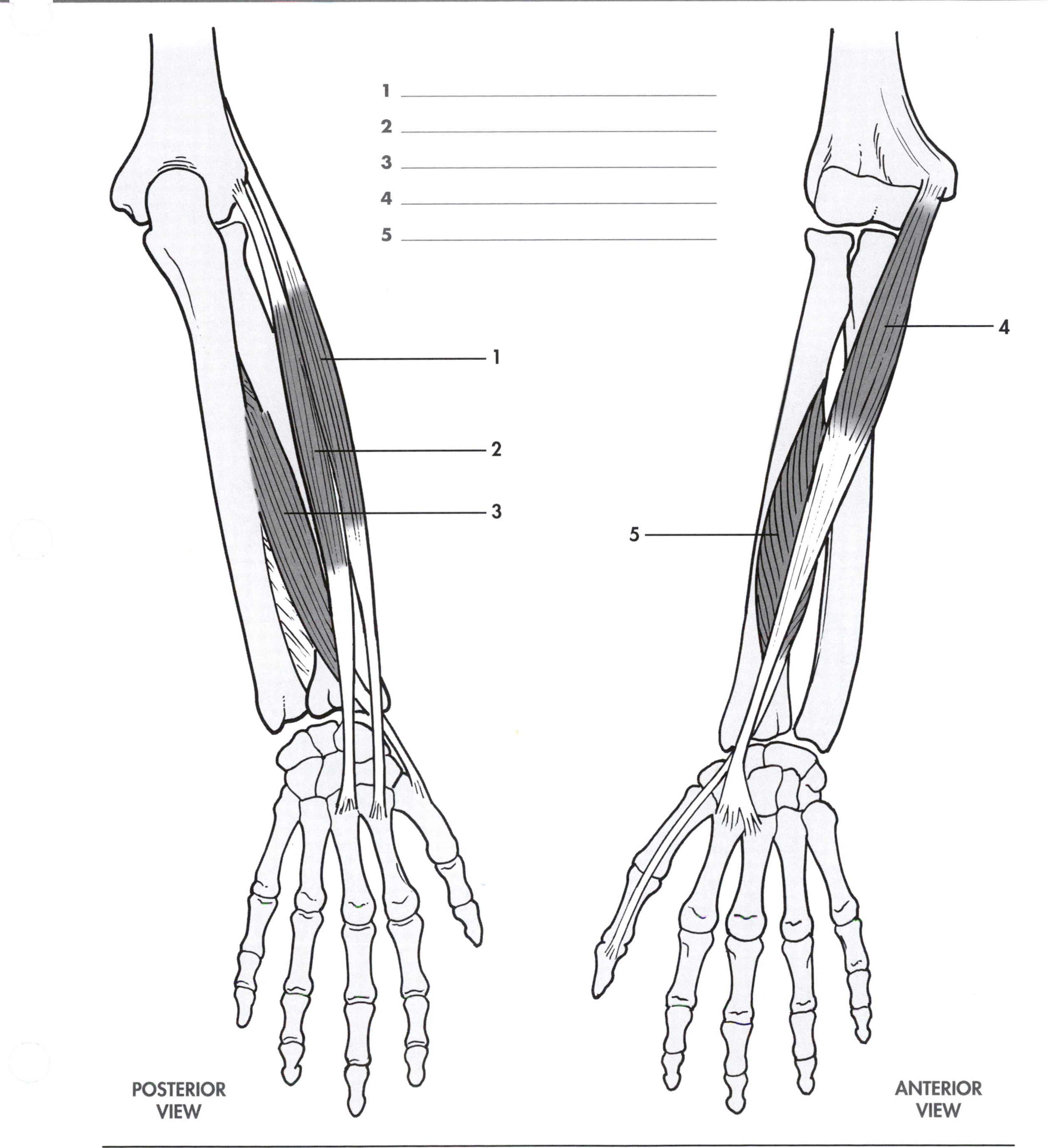

Abductors of Thumb and Digits

(**ab**•duck•tors)

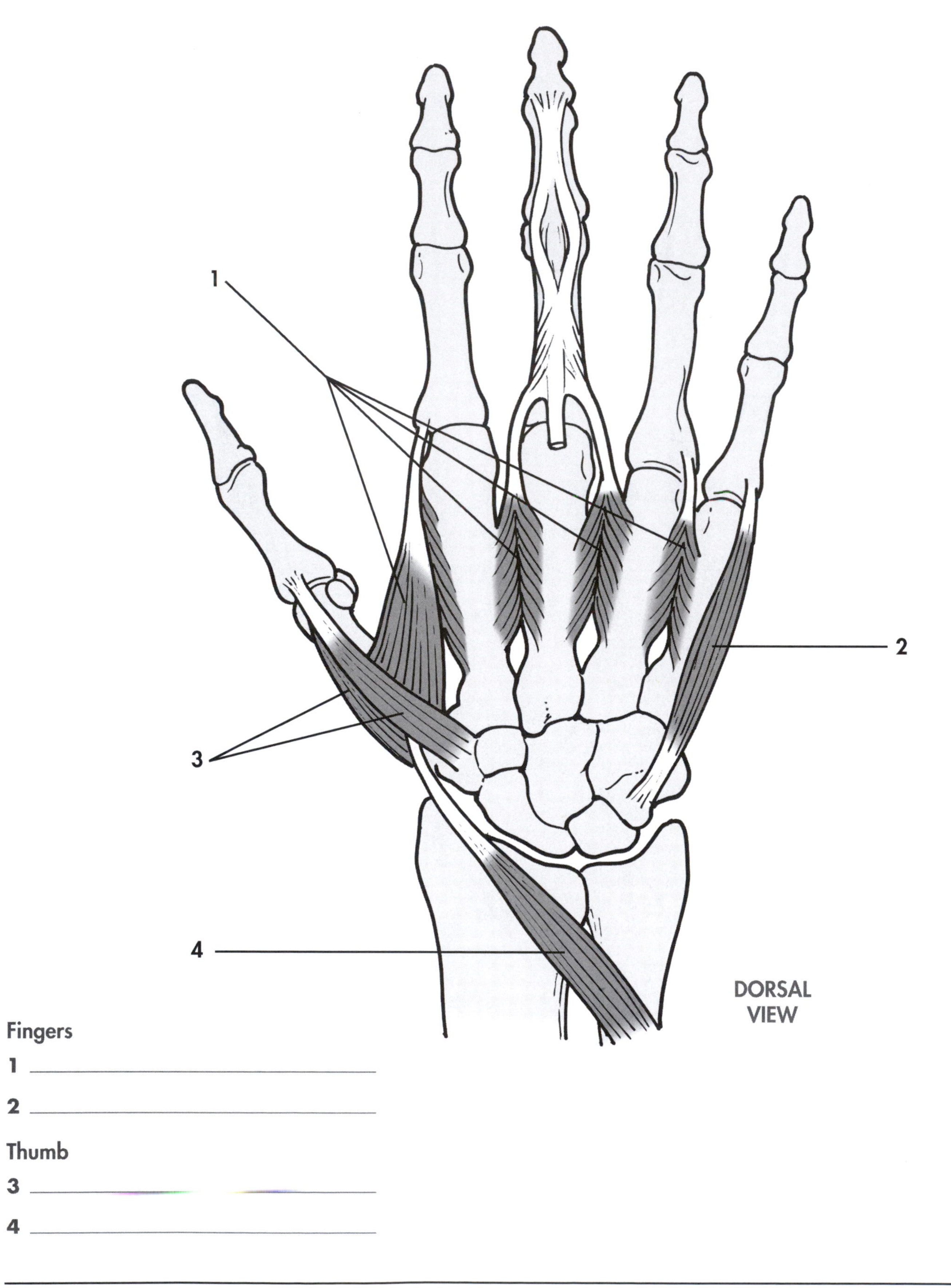

Fingers

1 ______________________________

2 ______________________________

Thumb

3 ______________________________

4 ______________________________

Adductors of Thumb and Digits

(**ad**•duck•tors)

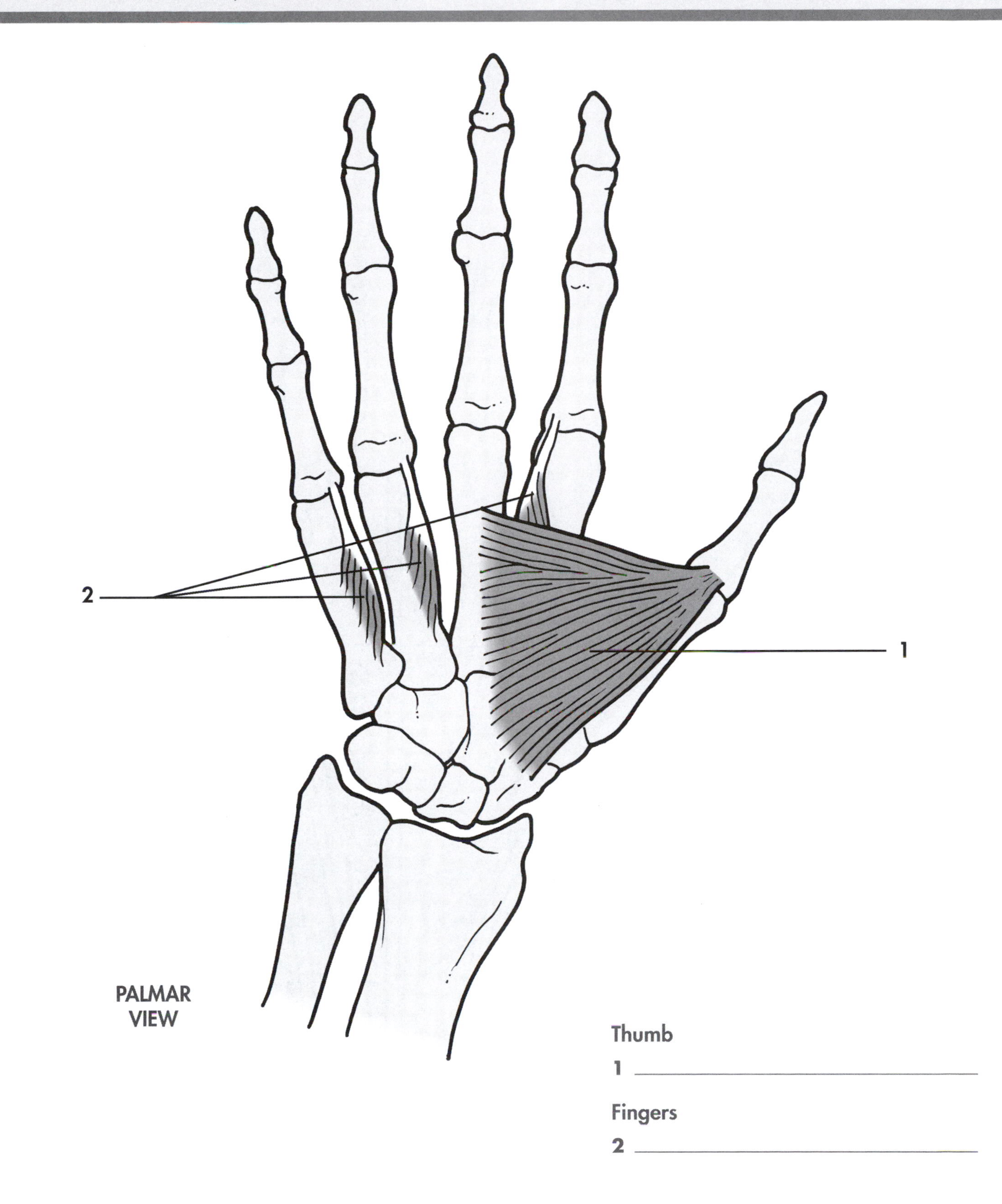

Extension of Thumb and Digits

(ex•**sten**•shun)

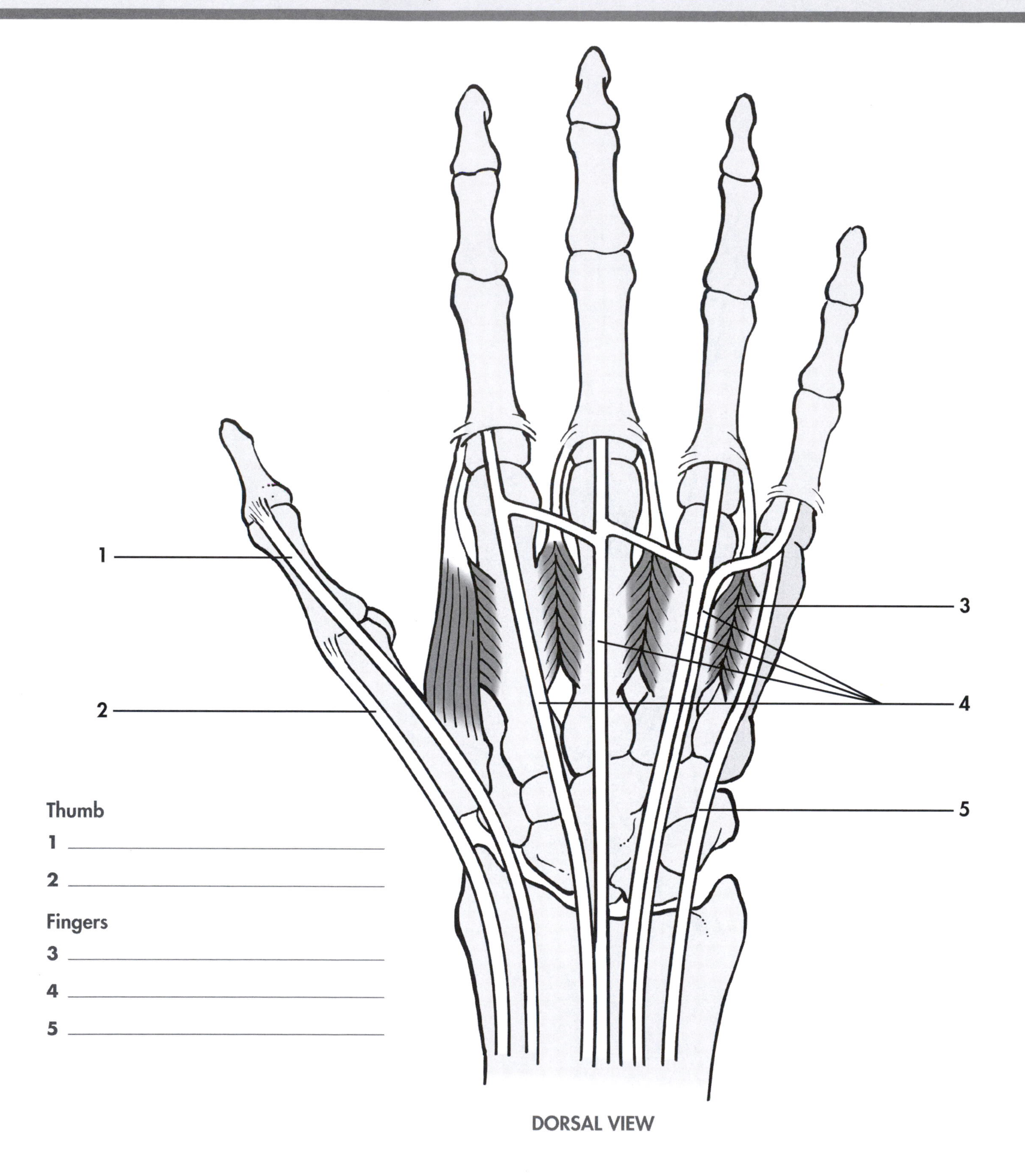

Flexion-Opposition of Thumb and Digits

(**flek**•shun) (**op**•poh•zih•shun)

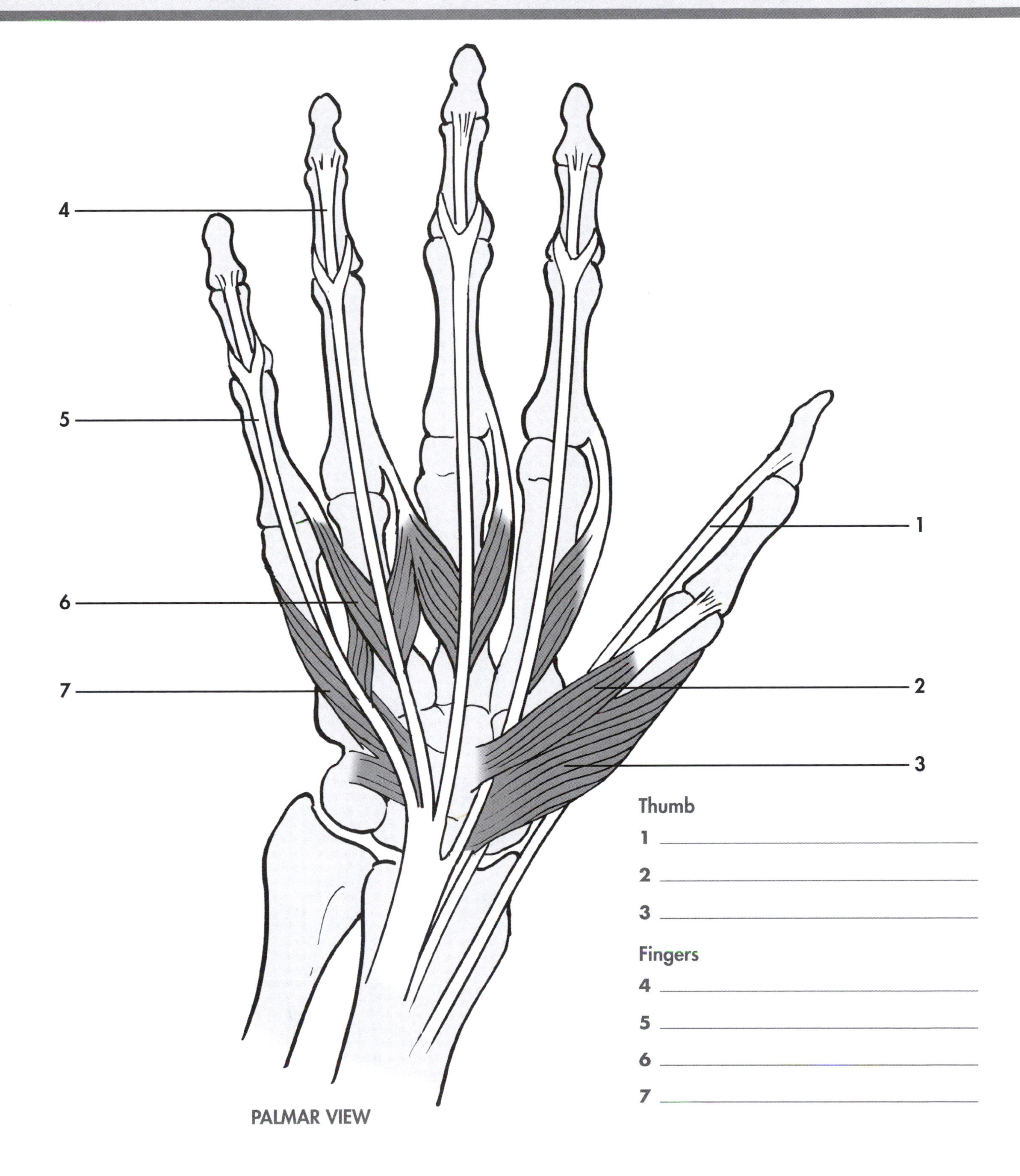

Lateral Rotators of Hip

(ro•**tay**•tors)

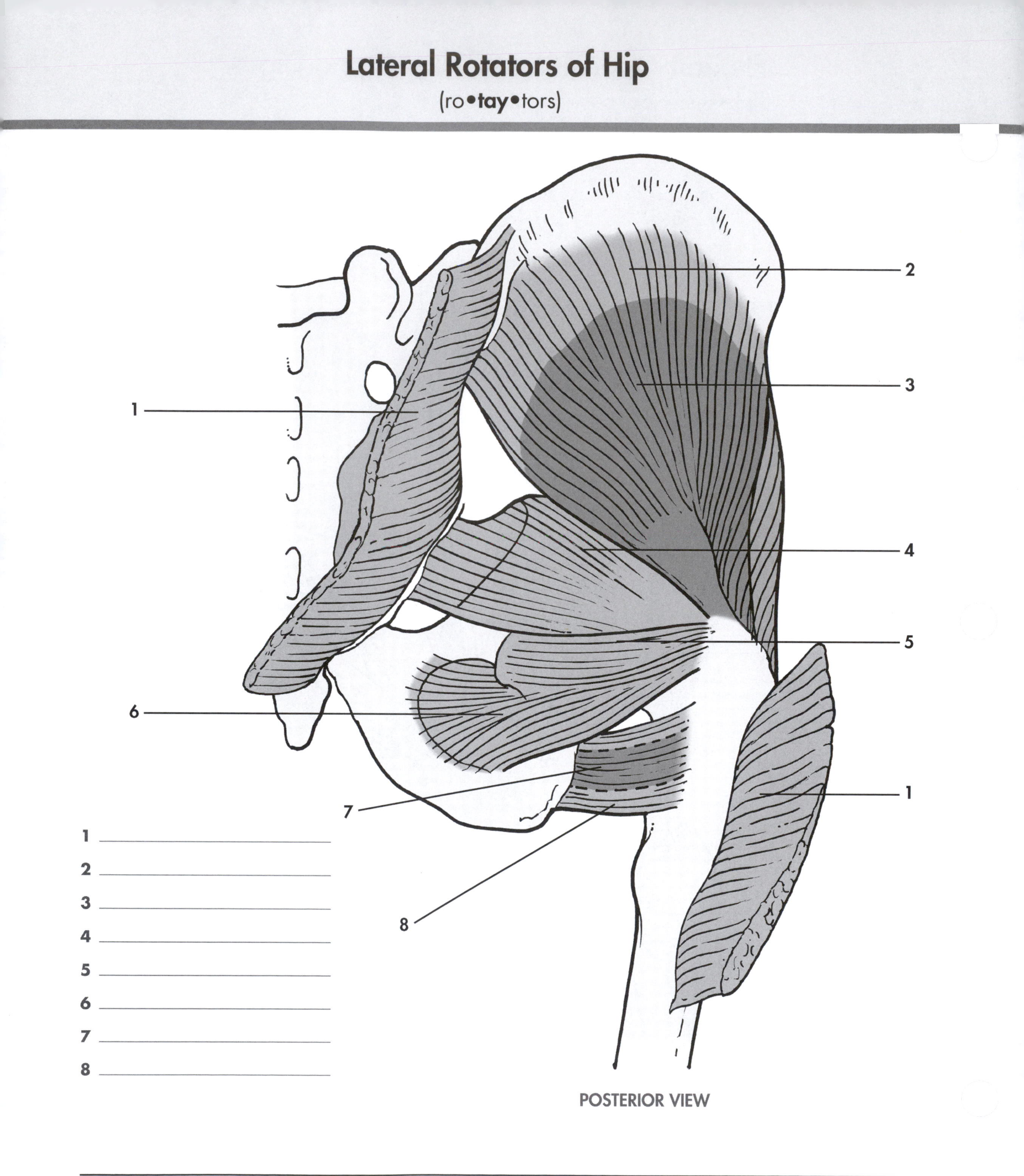

Medial Rotators of Hip

(ro•**tay**•tors)

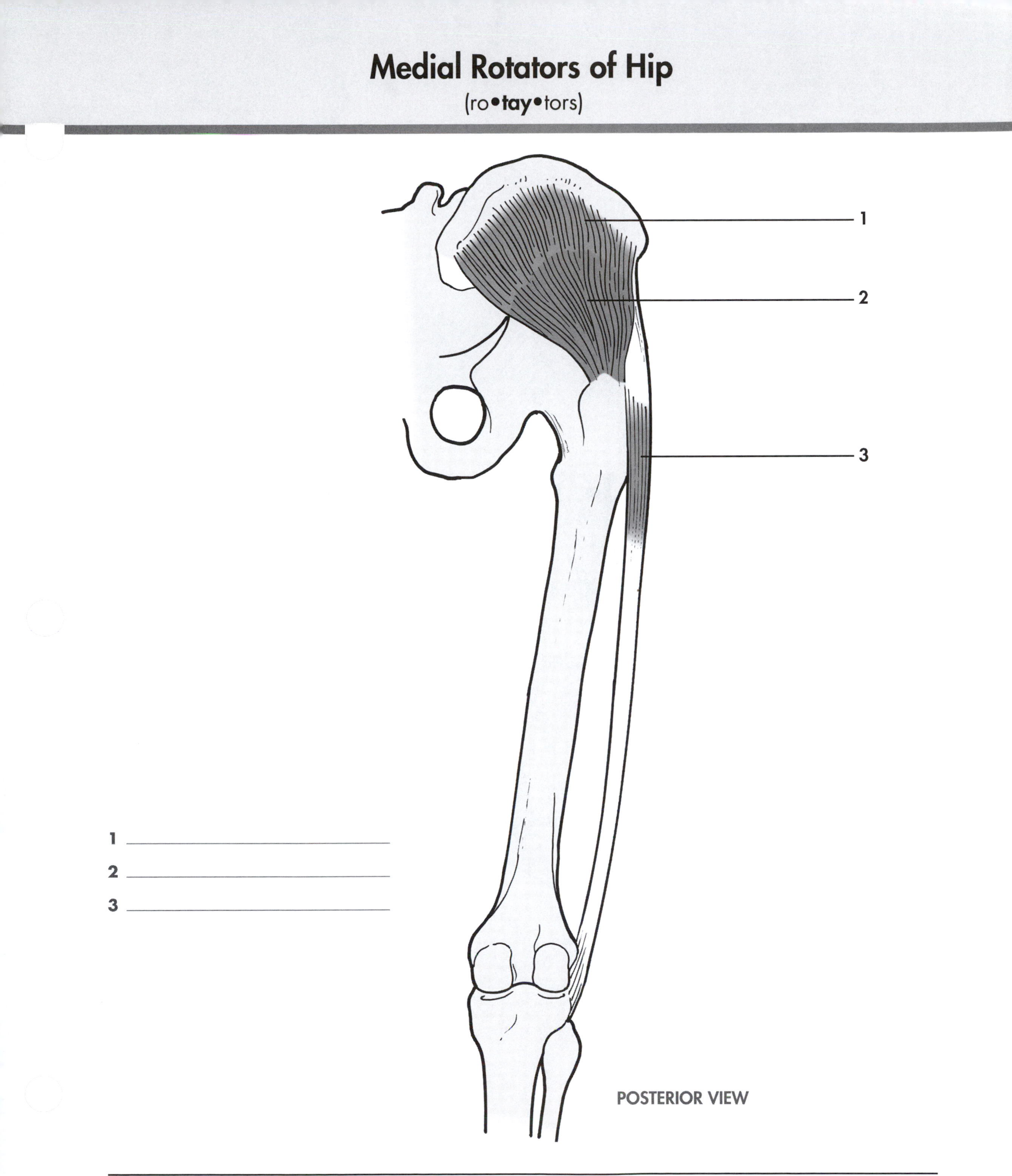

Extensors of Hip

(ex•**sten**•sors)

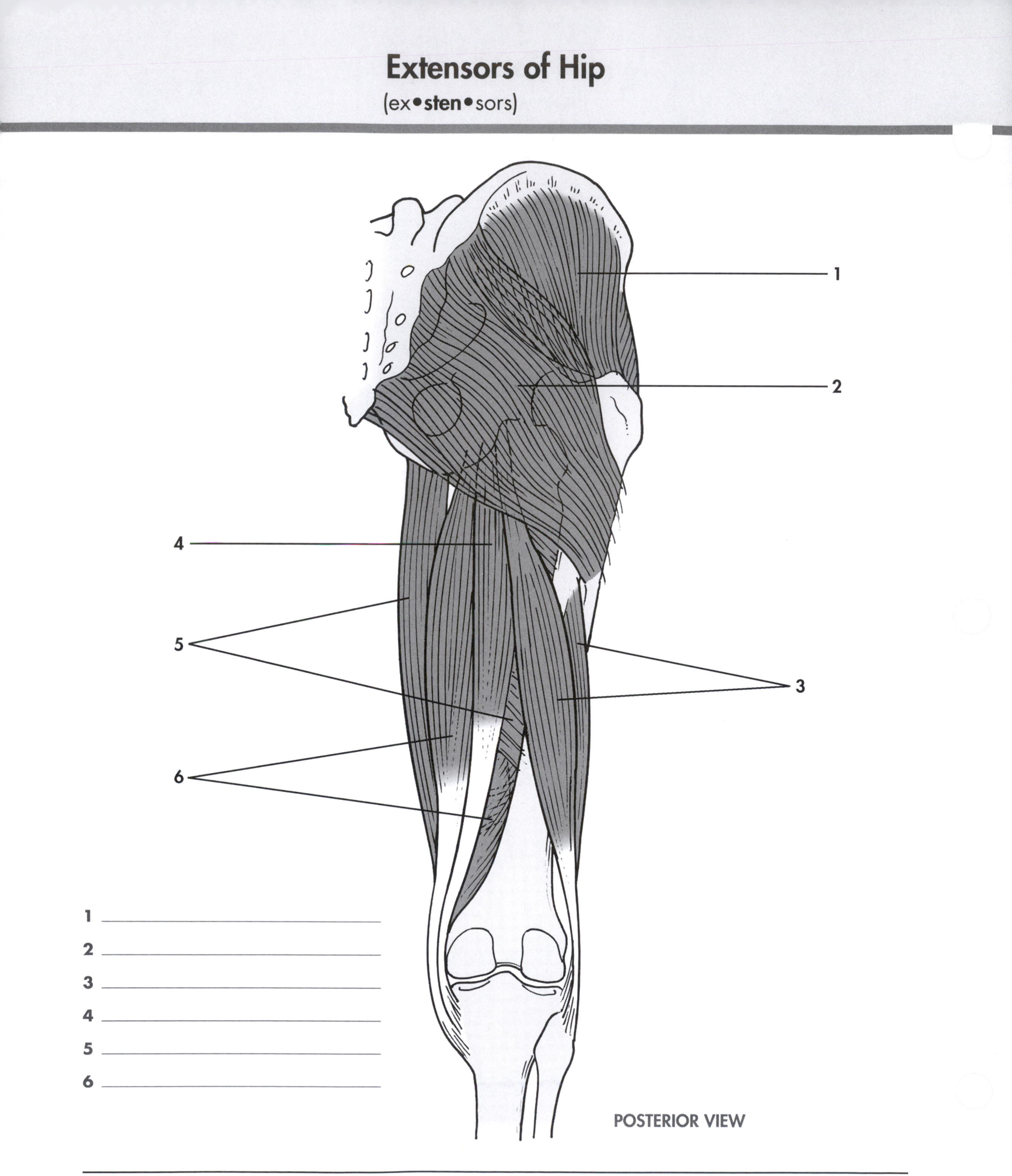

1 ______________________

2 ______________________

3 ______________________

4 ______________________

5 ______________________

6 ______________________

Flexors of Hip

(flek•sors)

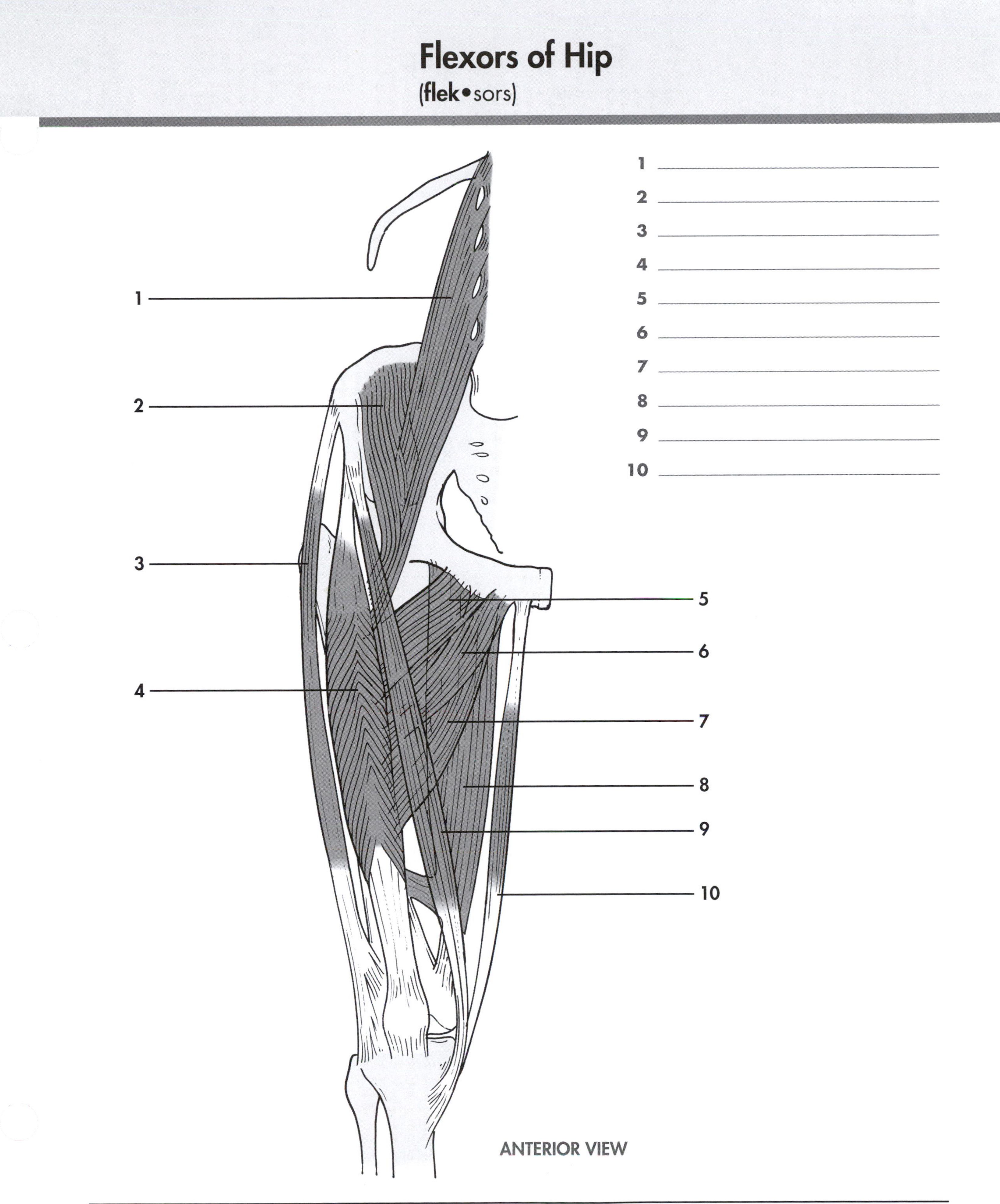

Abductors of Hip

(ab•duck•tors)

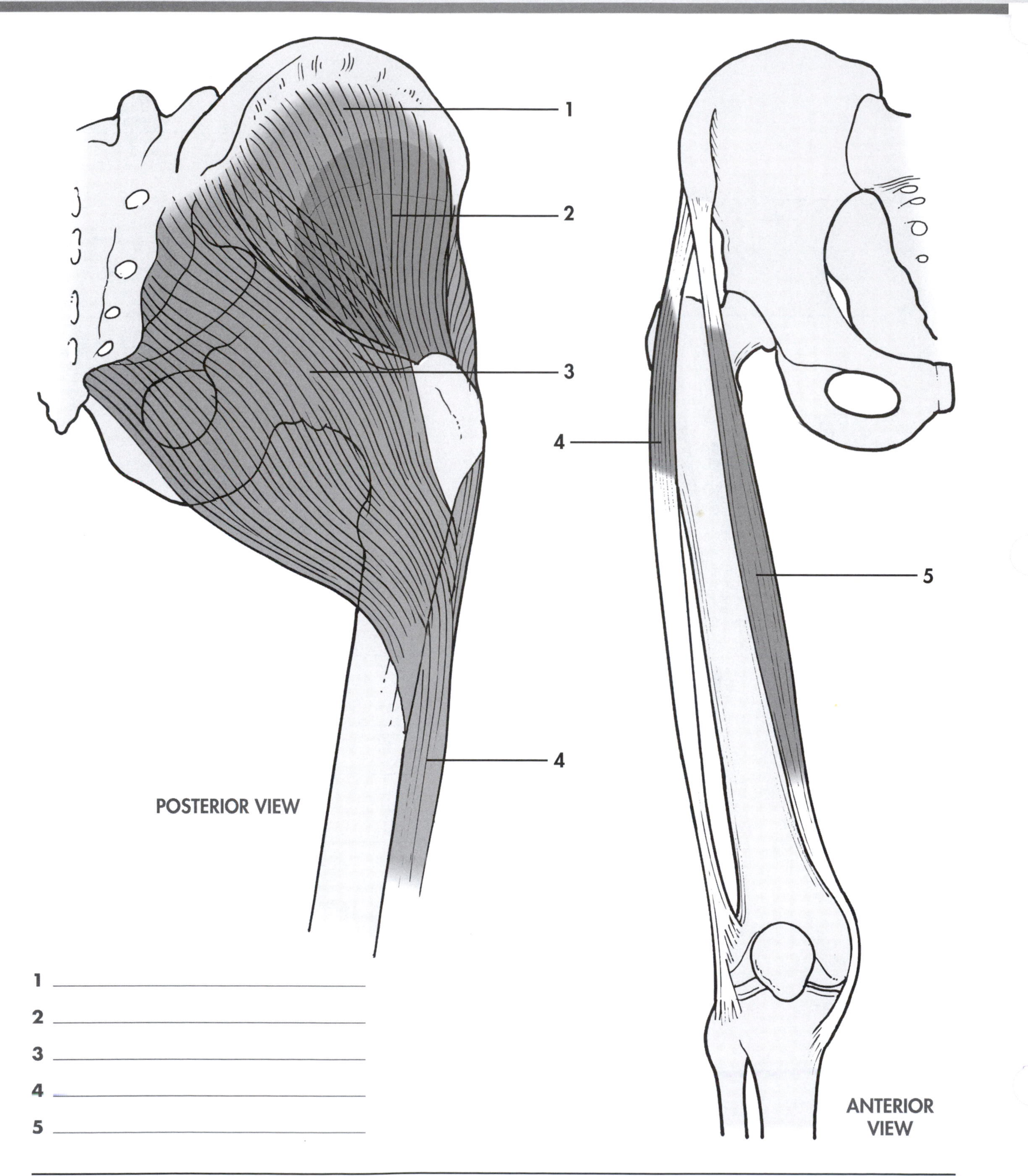

Adductors of Hip

(**ad**•duck•tors)

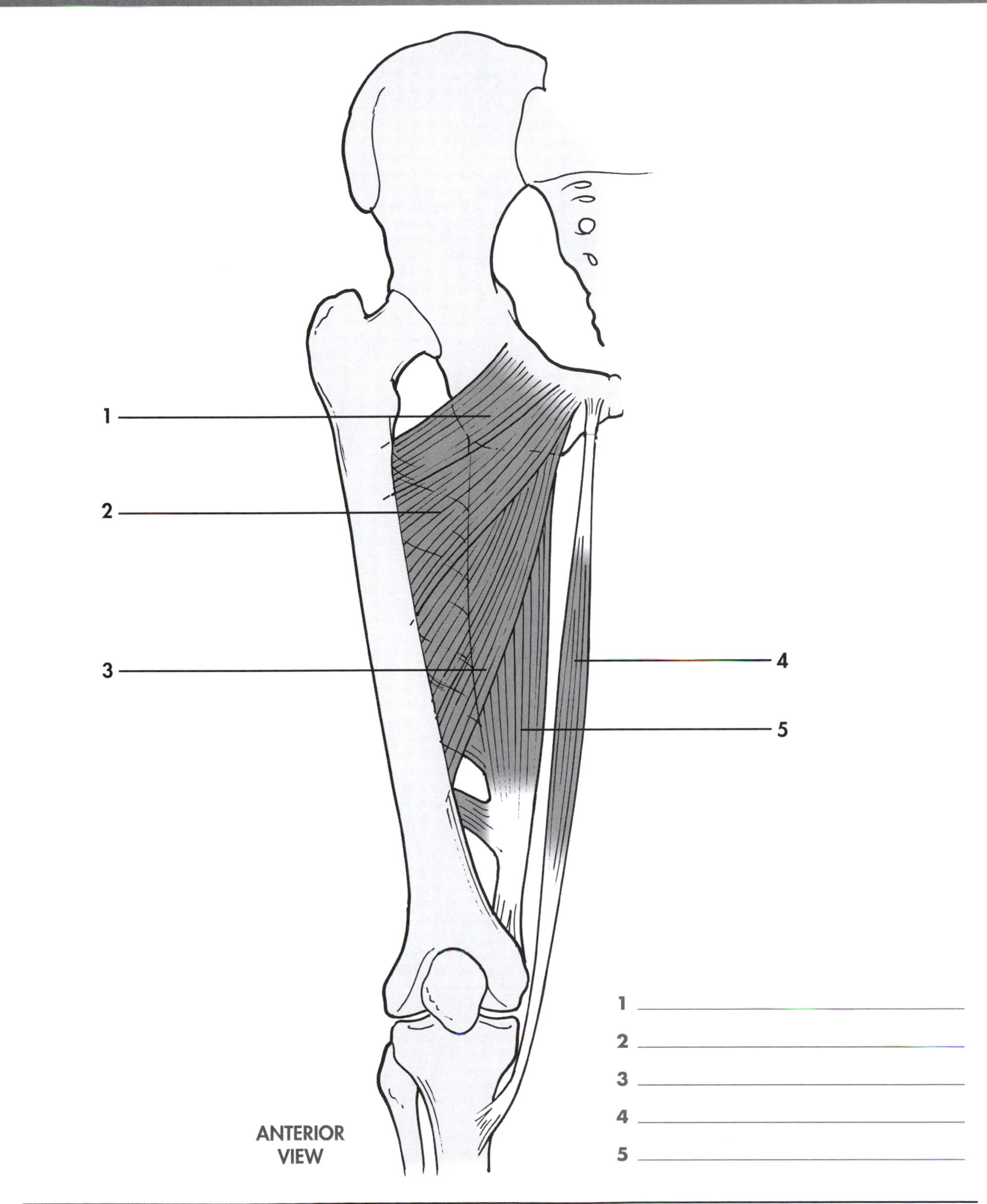

Lateral Rotators of Knee

(ro•**tay**•tors)

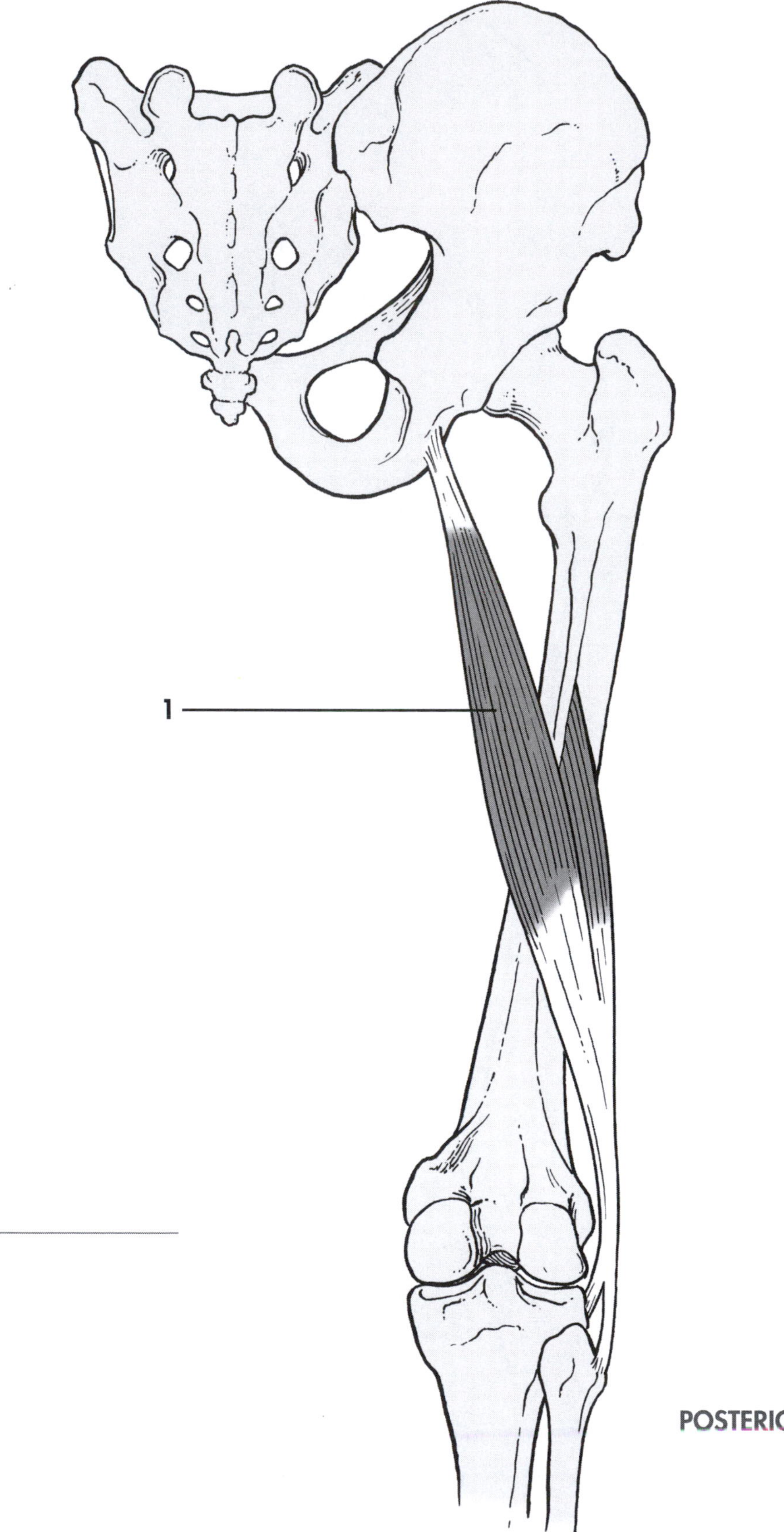

1 ______________________

Medial Rotators of Knee

(ro•**tay**•tors)

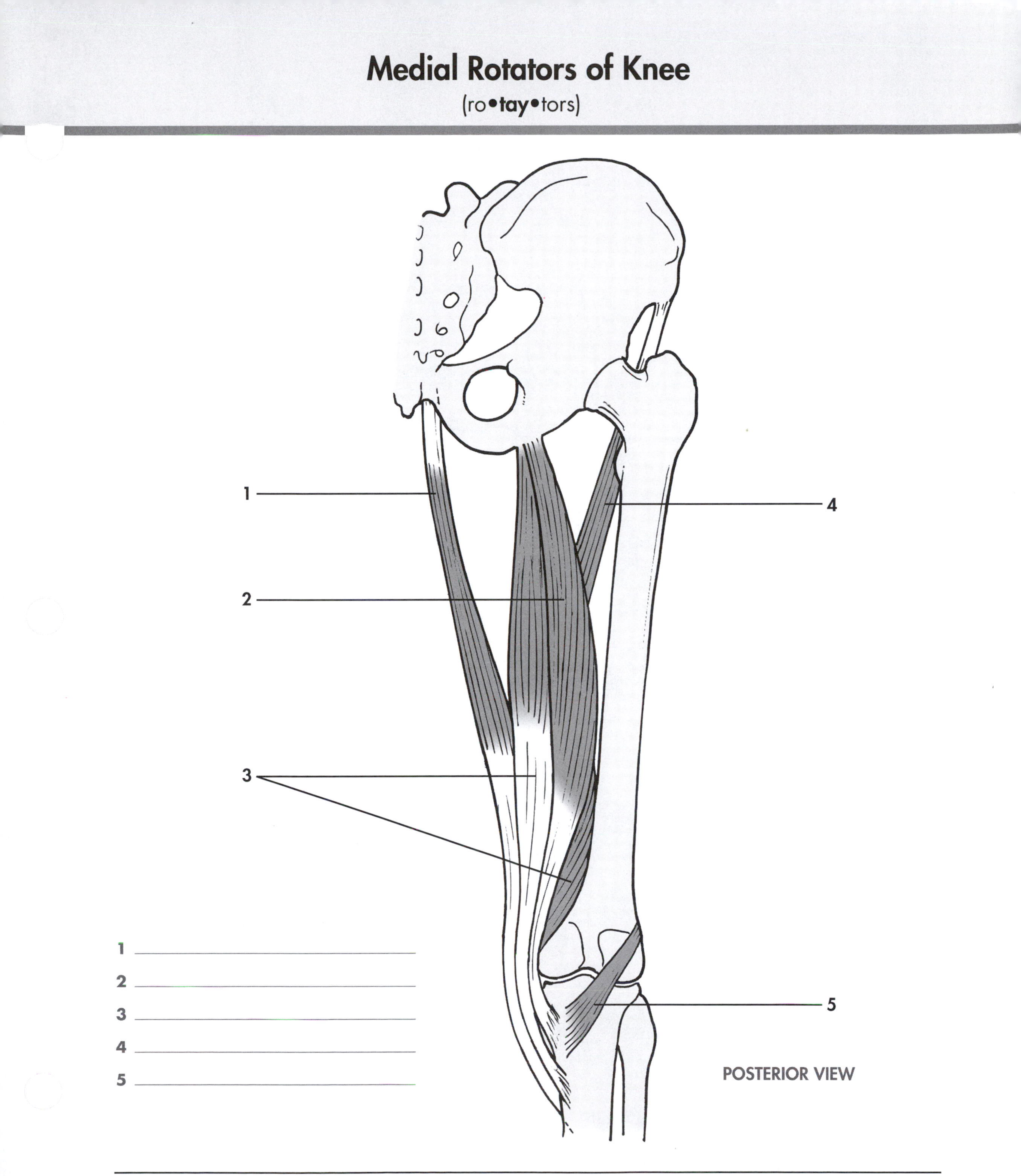

1 ______________________

2 ______________________

3 ______________________

4 ______________________

5 ______________________

Extensors of Knee

(ex•**sten**•sors)

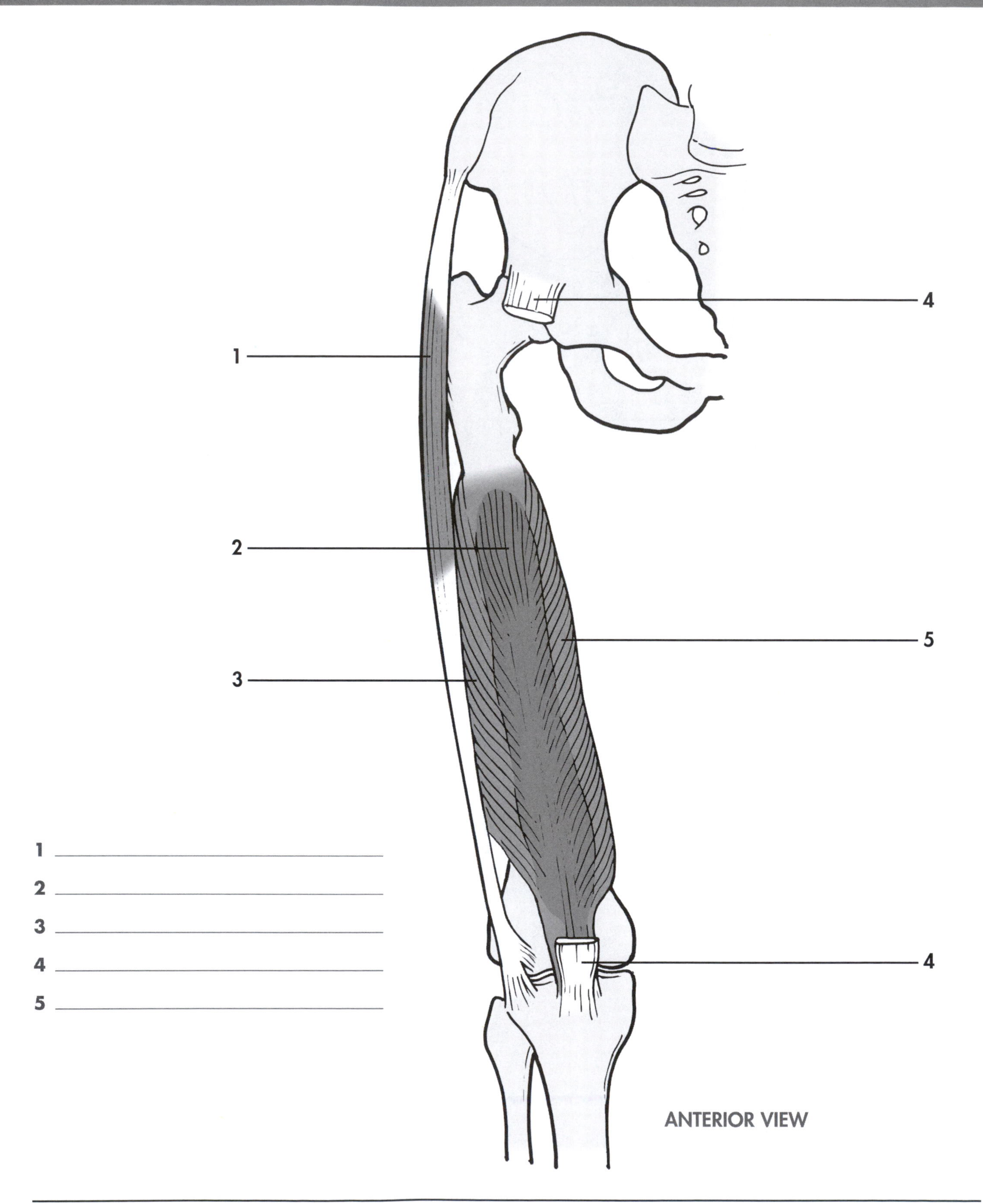

1 ______________________

2 ______________________

3 ______________________

4 ______________________

5 ______________________

Flexors of Knee

(flek•sors)

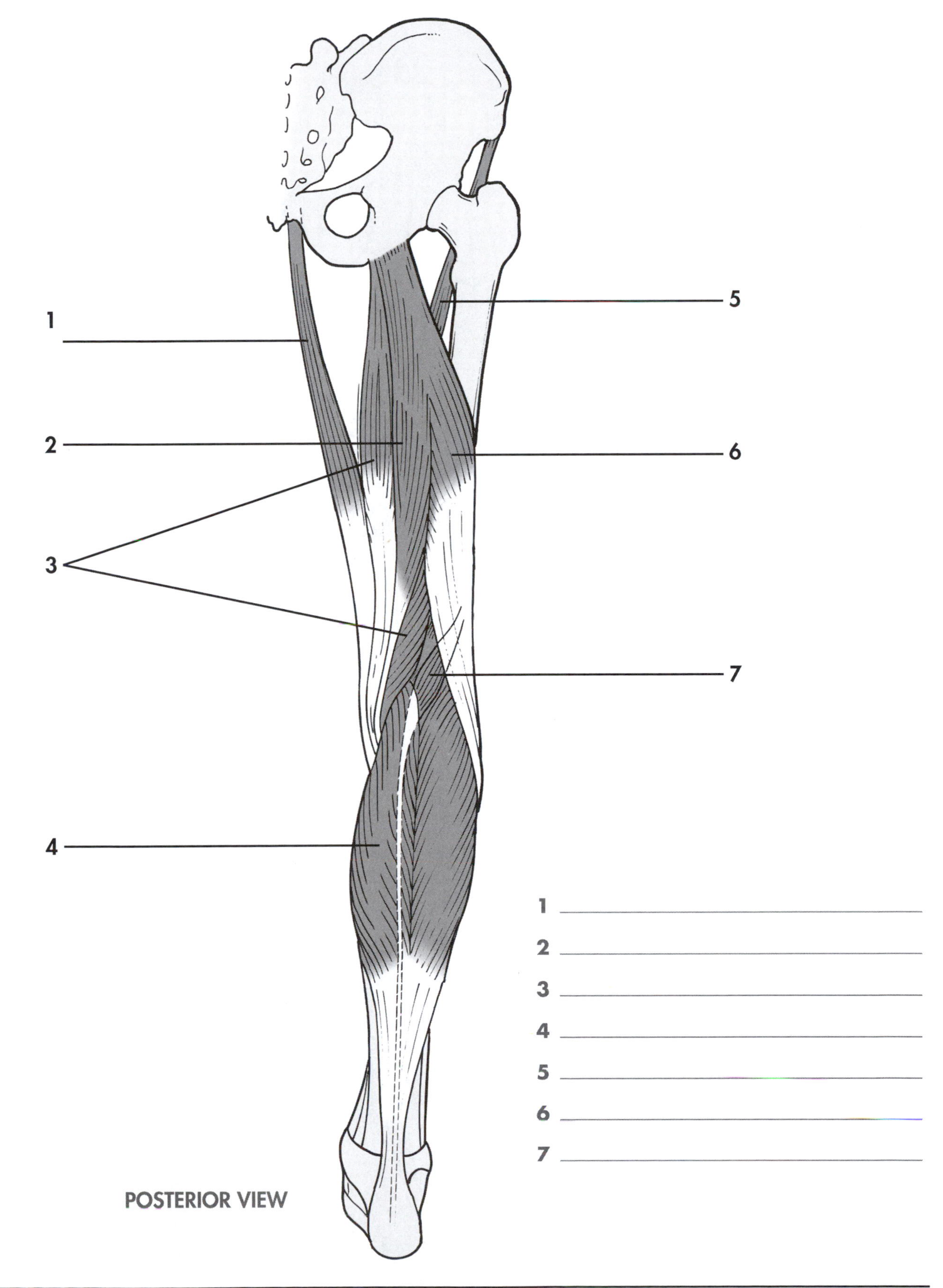

1 ______________________

2 ______________________

3 ______________________

4 ______________________

5 ______________________

6 ______________________

7 ______________________

Dorsiflexors of Ankle

(door•se•**flek**•sors)

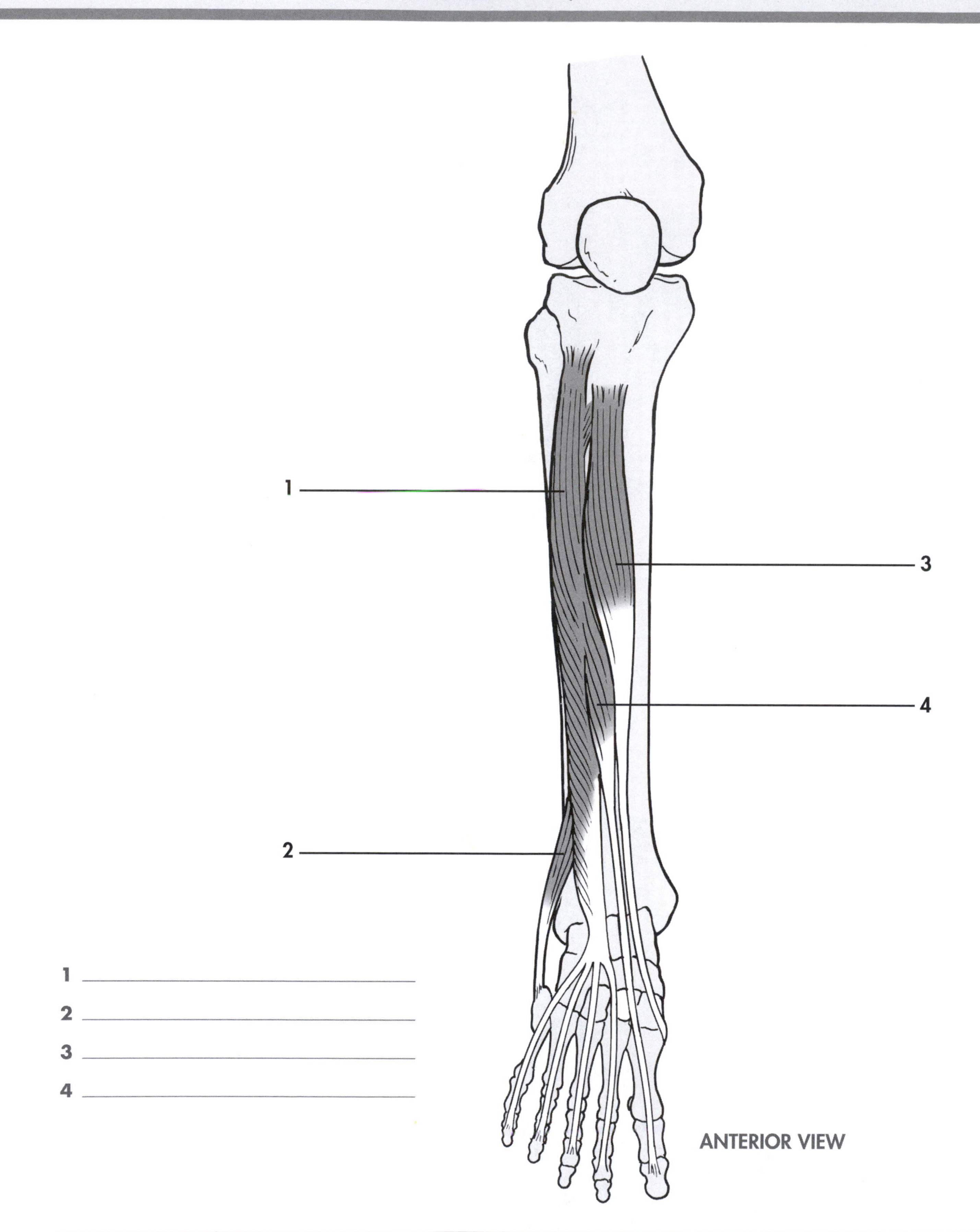

Plantar Flexors of Ankle

(plan•tahr) (flek•sors)

1 ______________________

2 ______________________

3 ______________________

4 ______________________

5 ______________________

6 ______________________

7 ______________________

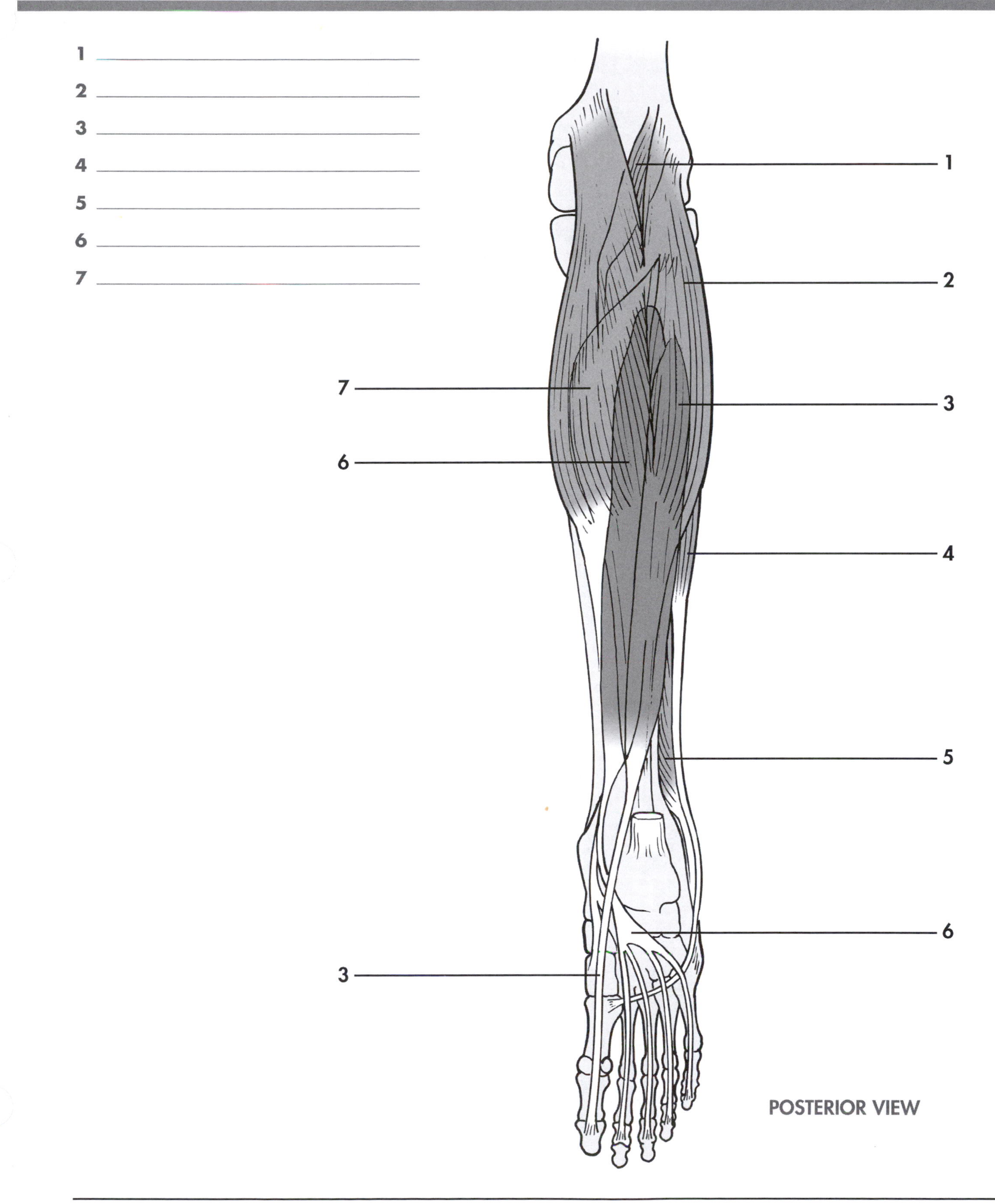

Invertors of Foot

(in•**ver**•tors)

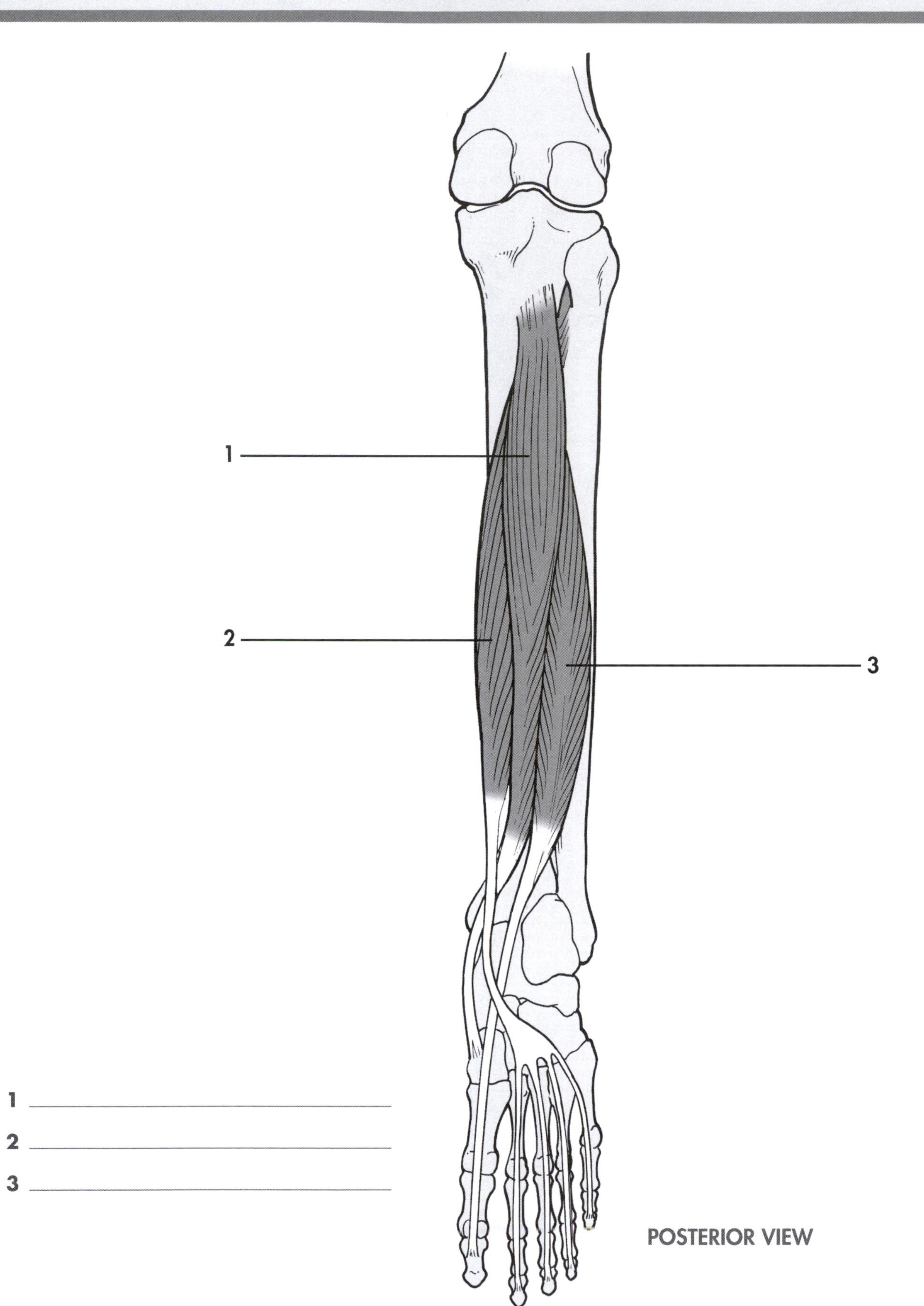

1 ______________________

2 ______________________

3 ______________________

Evertors of Foot

(ee•**ver**•tors)

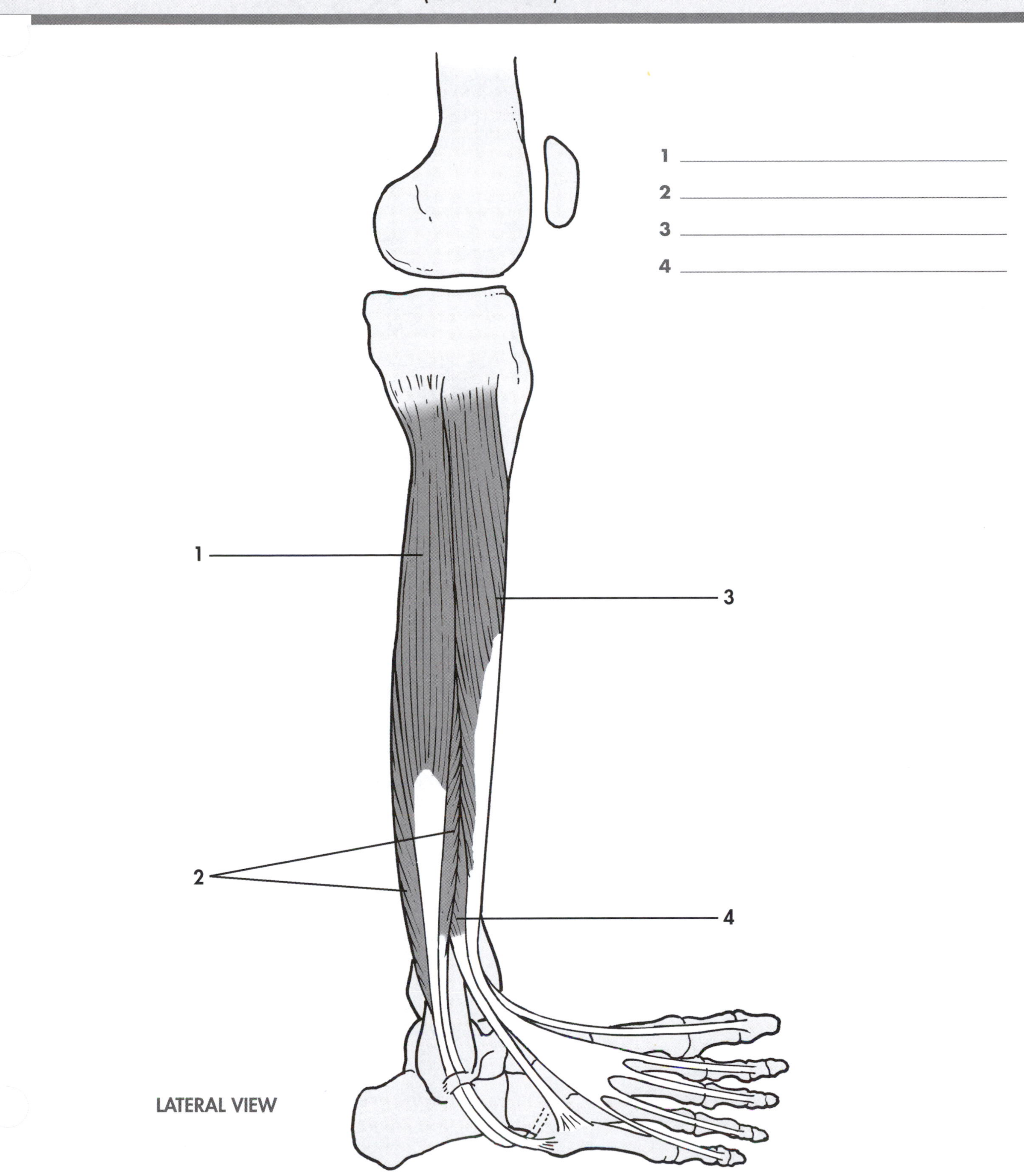

II. MATCHING

Group A: In the blank to the left of each muscle, write the **LETTER(S)** of the functional muscle group(s) in which each muscle participates. Each lettered functional muscle group may be used once, more than once, or not at all. Each muscle may belong in one or more than one functional muscle group.

1. _____ Pectoralis major
2. _____ Trapezius
3. _____ Biceps brachii
4. _____ Rhomboideus major
5. _____ Latissimus dorsi
6. _____ Teres major

A. Elevators of scapula
B. Depressors of scapula
C. Upward rotators of scapula
D. Retractors of scapula
E. Extensors of humerus
F. Medial rotators of humerus
G. Adductors of humerus
H. Flexors of elbow
I. Supinators of elbow

Group B: In the blank to the left of each muscle, write the **LETTER(S)** of the functional muscle group(s) in which each muscle participates. Each lettered functional muscle group may be used once, more than once, or not at all. Each muscle may belong in one or more than one muscle group.

1. _____ Gracilis
2. _____ Gastrocnemius
3. _____ Semimembranosus
4. _____ Adductor magnus
5. _____ Gluteus medius
6. _____ Gluteus maximus

A. Medial rotators of hip
B. Lateral rotators of hip
C. Extensor of hip
D. Flexor of hip
E. Adductors of hip
F. Medial rotators of knee
G. Flexors of knee
H. Plantar flexor of ankle

III. SENTENCE COMPLETION

Circle the term in parentheses that correctly completes each statement.

1. The depressors of the scapula *do not* include the (**latissimus dorsi, pectoralis minor, pectoralis major, serratus anterior**) muscle.
2. The upward rotators of the scapula *do not* include the (**serratus anterior, levator scapula, upper trapezius**) muscle.
3. The medial rotators of the humerus *do not* include the (**anterior deltoid, posterior deltoid, latissimus dorsi, subscapularis**) muscle.
4. The flexors of the humerus *do not* include the (**coracobrachialis, biceps brachii, brachialis, anterior deltoid**) muscle.
5. The adductors of the humerus *do not* include the (**teres major, teres minor, pectoralis major, anterior deltoid**) muscle.
6. Supinators of the forearm *do not* include the (**brachialis, brachioradialis, biceps brachii, supinator**) muscle.

7. Abductors of the wrist *do not* include the (**extensor carpi radialis longus, flexor carpi radialis, flexor pollicis longus, extensor carpi ulnaris**) muscle.

8. The medial rotators of the hip *do not* include the (**gluteus medius, gluteus minimus, sartorius, tensor fascia latae**) muscle.

9. Flexors of the hip *do not* include the (**sartorius, gracilis, semitendinosus, rectus femoris, psoas major**) muscle.

10. Medial rotators of the knee *do not* include the (**semitendinosus, semimembranosus, biceps femoris, gracilis, popliteus**) muscle.

11. Evertors of the foot *do not* include the (**fibularis longus, fibularis brevis, tibialis anterior, extensor digitorum longus**).

12. Dorsiflexors of the ankle *do not* include the (**extensor digitorum longus, tibialis anterior, fibularis longus, fibularis tertius**).

13. Plantar flexors of the ankle *do not* include the (**plantaris, gastrocnemius, fibularis longus, extensor hallucis longus**).

14. Extensors of the knee *do not* include the (**tensor fascia latae, rectus femoris, gastrocnemius, vastus lateralis**).

15. Adductors of the hip *do not* include the (**sartorius, gracilis, adductor longus, pectineus**).

16. The origin of the adductors of the hip is on the (**ischium, ilium, femur, pubis**) bone.

17. The origin of the medial rotators of the hip is on the (**ischium, ilium, femur, pubis**) bone.

18. The insertion of the abductors of the wrist is on the (**carpals, ulna, metacarpals, phalanges**).

19. The insertion of the extensors of the elbow is on the (**olecranon process, radial tuberosity, epicondyle of the humerus, styloid process of ulna**).

20. The origin of the lateral rotators of the humerus is on the (**clavicle, vertebrae, scapula, sternum**).

IV. MULTIPLE CHOICE

Write the LETTER of the correct answer in the blank to the left of each question.

1. _____ Which of the following muscles *is not* part of the quadriceps group?
   A. Vastus lateralis
   B. Rectus femoris
   C. Biceps femoris
   D. Vastus medialis

2. _____ All of the following are invertors of the foot *except* the
   A. Tibialis posterior
   B. Flexor hallucis longus
   C. Flexor digitorum longus
   D. Fibularis longus

3. _____ Simultaneous contraction of the gastrocnemius, soleus, and plantaris muscles causes
   A. Dorsiflexion of the ankle
   B. Plantar flexion of the ankle
   C. Inversion of the foot
   D. Eversion of the foot

4. _____ Simultaneous contraction of the hamstring group and the gastrocnemius would cause
   A. Flexion of the knee
   B. Extension of the knee
   C. Flexion of the hip
   D. Extension of the hip

5. ______ Simultaneous contraction of the three gluteal muscles, the sartorius, and tensor fascia latae, would cause
   A. Adduction of the hip
   B. Flexion of the hip
   C. Abduction of the hip
   D. Medical rotation of the hip

6. ______ Simultaneous contraction of the extensor carpi radialis longus and brevis and the flexor carpi radialis muscles will cause
   A. Flexion of the wrist
   B. Extension of the wrist
   C. Abduction of the wrist
   D. Adduction of the wrist

7. ______ The major extensors of the elbow are the
   A. Brachioradialis and supinator
   B. Triceps brachii and anconeus
   C. Triceps brachii and supinator
   D. Brachioradialis and anconeus

8. ______ The latissimus dorsi and teres major are two posterior muscles that ________ the humerus.
   A. Flex
   B. Laterally rotate
   C. Abduct
   D. Adduct

9. ______ All of the following are rotator cuff muscles *except* the
   A. Subscapularis
   B. Infraspinatus
   C. Teres minor
   D. Teres major

10. ______ The trapezius participates in all of the following actions *except* the
    A. Elevation of the scapula
    B. Medial rotation of the humerus
    C. Upward rotation of the scapula
    D. Retraction of the scapula

11. ______ All of the following are elevators of the scapula *except* the
    A. Upper trapezius
    B. Rhomboideus major
    C. Latissimus dorsi
    D. Levator scapulae

12. ______ The serratus anterior, pectoralis major, and pectoralis minor work together to ________ the scapula.
    A. Depress
    B. Retract
    C. Protract
    D. Upwardly rotate

13. ______ Lateral rotators of the humerus include all of the following *except* the
    A. Infraspinatus
    B. Supraspinatus
    C. Posterior deltoid
    D. Teres minor

14. ______ The anterior deltoid, coracobrachialis, and biceps brachii are all ________ of the humerus
    A. Extensors
    B. Flexors
    C. Adductors
    D. Abductors

15. ______ All of the following are extensors of the wrist *except* the
    A. Extensor carpi radialis longus
    B. Extensor digitorum
    C. Extensor pollicis longus
    D. Extensor carpi ulnaris

16. ______ The fingers are abducted by the
    A. Dorsal interossei
    B. Palmar interossei
    C. Extensor digitorum
    D. Opponens digiti

17. _____ All of the following muscles flex digits 2–5 *except* the
   A. Flexor pollicis
   B. Lumbricales
   C. Flexor digitorum superficialis
   D. Palmar interossei

18. _____ Simultaneous contraction of the gluteus medius, gluteus minimus, and tensor fascia latae, would cause
   A. Lateral rotation of hip
   B. Adduction of hip
   C. Medial rotation of hip
   D. Extension of hip

19. _____ Many muscles work together to flex the hip. All of the following are flexors of the hip *except* the
   A. Psoas major
   B. Rectus femoris
   C. Biceps femoris
   D. Tensor fascia latae

20. _____ The tensor fascia latae performs many actions and is included in several functional muscle groups. It is part of all of the following muscle groups *except* the
   A. Abductors of the hip
   B. Flexors of the hip
   C. Lateral rotators of hip
   D. Extensors of the knee

V. FILL IN THE BLANK

Write the correct word(s) in the blank spaces to complete each statement.

1. Name a muscle that belongs to both the dorsiflexors of the ankle and invertors of the foot ________________ .
2. The extensors of the knee include the four quadriceps muscles and the ________________ .
3. The three hamstrings muscles are the ________________ , ________________ , and ________________ .
4. The most lateral of the hamstrings and the one that is a lateral rotator of the knee is the ________________ .
5. The muscle that abducts the hip and goes from the anterior superior iliac spine to the proximal medial tibia is the ________________ .
6. The gluteus muscle that does *not* extend the hip is the ________________ .
7. The muscle in the hand that allows us to bring our thumb to the little finger is the ________________ .
8. Two muscles that adduct the wrist are the ________________ and the ________________ .
9. The palmaris longus is part of what functional muscle group? ________________
10. The extensors of the elbow are the triceps brachii and the small ________________ .
11. Disease- or injury-induced paralysis of the ________________ muscle group would prevent straightening the lower leg.

12. Protractors of the scapula include the __________________ and __________________ muscles.

13. The subscapularis is part of what functional muscle group? __________________

14. The rhomboideus major and minor are both part of the __________________ and __________________ functional muscle groups.

15. The flexors of the elbow include the __________________, __________________, __________________, and __________________.

16. The two small flexors of the knee that are found in the popliteal fossa are the __________________ and __________________.

17. The muscles that allow adduction of the fingers are the __________________.

18. The extensor of the wrist muscles insert on the __________________ surface of the hand.

19. The insertion of the __________________ muscle splits to allow the insertion of the __________________ muscle to pass through on its way to the terminal phalanges.

20. The small muscle that passes between the distal ends of the radius and ulna to pronate the forearm is the __________________.

VI. SHORT ANSWER

1. Rather than just memorize a muscle name, *learn the meaning of the name* because it provides useful information on one or more of the following characteristics: location, size, shape, direction of muscle fibers, action, and attachments. For each of the following muscles, indicate what specific information each portion of the muscle name provides.

   A. Flexor digitorum ______________________________

   B. Temporalis ______________________________

   C. Biceps brachii ______________________________

D. Deltoid ______________________________

____

E. External oblique ______________________________

____

F. Adductor magnus ______________________________

____

G. Coracobrachialis ______________________________

____

2. Write the correct spelling of each term in the blank.

   A. sinergistic ______________________________

   B. tendin ______________________________

   C. latisimus ______________________________

   D. gastronemius ______________________________

   E. radiel ______________________________

   F. glooteus ______________________________

3. Although abbreviated slang such as "pecs," "biceps," "abs," and "rhoms" may be used in a training room or a study group, why is it best to use the full names of these muscles, particularly in a clinical setting and on written and practical examinations?

____

____

____

VII. FILL IN THE CHART

Write the **specific points of attachment** for the individual muscles that participate in the designated functional muscle groups.

MUSCLE	ORIGIN	INSERTION
Flexors of the Elbow		
Biceps brachii		
Brachialis		
Brachioradialis		
Pronator teres		
Extensors of the Elbow		
Triceps brachii		
Anconeus		

MUSCLE	ORIGIN	INSERTION
Supinators of the Forearm		
Supinator		
Biceps brachii		
Brachioradialis		
Pronators of the Forearm		
Pronator teres		
Pronator quadratus		
Brachioradialis		

MUSCLE	ORIGIN	INSERTION
Flexors of the Knee		
Gracilis		
Semitendinosus		
Semimembranosus		
Gastrocnemius		
Sartorius		
Biceps femoris		
Plantaris		

MUSCLE	ORIGIN	INSERTION
Extensors of the Knee		
Tensor fascia latae		
Vastus lateralis		
Vastus medialis		
Vastus intermedius		
Rectus femoris		

MUSCLE	ORIGIN	INSERTION
Plantar Flexors of the Ankle		
Plantaris		
Gastrocnemius		
Soleus		
Fibularis longus		
Fibularis brevis		
Flexor digitorum longus		

MUSCLE	ORIGIN	INSERTION
Dorsiflexors of the Ankle		
Extensor digitorum longus		
Fibularis tertius		
Tibialis anterior		
Extensor hallucis longus		

VIII. CASE STUDIES

A. A 12-year-old male is brought to the emergency room having suffered an injury in a soccer game. He describes feeling a severe pain in the front knee region immediately after rapidly swinging his right lower leg forward to kick the ball. He states that he is not able to "straighten" his lower leg. Imaging shows that the thigh muscle tendon is not torn but that the "bony landmark" on the proximal, anterior surface of the "shin bone" to which the tendon attaches by way of the patella has separated from the lower leg bone.

1. Name the thigh muscle group he used to "bend" his lower leg back. ____________________

   What is the technical term for that motion? ____________________

2. Name the thigh muscle group he used to bring his lower leg forward to kick the soccer ball. ____________________

   What is the technical term for that motion? ____________________

3. Name the large lower leg bone on which both muscle groups insert. ____________________

4. Name the specific "bony landmark" that separated from his lower leg bone. ____________________

5. Name the specific "youth sports injury" that the youngster has incurred. ____________________

6. Explain why the "bony landmark" readily separated from the main bone in reaction to the force of the original ball kick by this 12-year-old youngster but would not in a 25-year-old?

   ____

   ____

   ____

B. After a car accident, a 60-year-old man complained of pain in his right hip. Examination showed that his leg was adducted and medially rotated, and his knee was slightly flexed. X-rays revealed posterior dislocation of the hip.

1. What is the name of the socket into which the head of the femur fits? ____________________

2. Why would this dislocation cause adduction of the leg?

   ____

   ____

   ____

3. What functional muscle group would be involved based on the symptoms? ____________________

4. Name some of the flexors of the knee.

____

____

5. Name the medial rotators of the hip.

____

____

6. Which nerve lies close to the hip joint, making it vulnerable to injury during dislocation?

____

IX. PALPATION EXERCISE

Now that you have reviewed the attachments and actions of various functional muscle groups, locate and palpate the following antagonistic sets of muscles as described below. This is best done with a partner. Palpation instructions are directed to the examiner.

FLEXION AND EXTENSION OF THE LOWER ARM

The examiner and subject sit opposite one another. The subject sits with the arm supinated on the thigh.

- Enhance the subject's contraction of the biceps brachii by resisting the flexion of the lower arm. Feel the **biceps brachii** between your thumb and index finger on the anterior surface of the upper arm. Palpate proximally and distally along its length. At the proximal end, the belly of the muscle is formed by the union of the **long head** from the **supraglenoid tubercle** laterally and the **short head** from the **coracoid process** medially. Follow the short head to the coracoid process. It is not practical to locate the supraglenoid tubercle.
- Enhance the subject's contraction of the triceps brachii by resisting the extension of the lower arm. Palpate the width of the **triceps brachii** from its insertion on the olecranon process up along the length of the back of the upper arm. Both the **long** and **lateral heads** are superficial except for an upper portion of each head, which lies deep to the posterior deltoid. Palpate each along the proximal two-thirds of the medial and lateral sides of the arm. Although the **medial head** is mostly beneath the other two heads, it can be palpated distally just above the medial and lateral epicondyles of the humerus.

PRONATION AND SUPINATION OF THE LOWER ARM

The examiner and subject sit opposite one another. The subject sits with the arm relaxed on the thigh with the forearm flexed halfway between pronation and supination. The examiner places his or her left hand on the subject's anterior proximal forearm and supports the subject's forearm just proximal to the wrist.

- The examiner resists the subject from fully pronating the forearm with the left hand and feels with the thumb of the right hand for the contraction of the **pronator teres**. It arises from the **medial epicondyle** and **supracondylar ridge of the humerus** and **coronoid process of the ulna** and runs diagonally to insert on the midlateral radius.
- The **pronator quadratus** is palpated by the left hand, holding the forearm firmly to resist full pronation of the forearm, and two fingers of the right hand firmly palpate the transversely placed muscle on the radial side at the anterior distal forearm. When contraction of the **pronator quadratus** is felt, follow the muscle across toward its ulnar attachment.

- To palpate the **supinator**, separate the radial group of muscles from the other muscles of the forearm by pushing the muscles aside with the index and middle fingers of the right hand. Gently press the fingers deeper toward the supinator attachment on the radius. Resist full supination of the hand and feel the contraction of the supinator. Palpate the supinator toward its proximal attachment and feel for its contraction as the subject alternately contracts and relaxes the supinator.

FLEXION AND EXTENSION OF THE LOWER LEG

- Have the subject sit with thighs on a bench and the legs hanging down. The great bulk of muscle on the front of the thigh is the quadriceps femoris group, which include the **vastus medialis, vastus intermedius, vastus lateralis**, and **rectus femoris.** All four converge to insert on the **tibial tuberosity** via the **patella ligament**. Except for the deep **vastus intermedius**, the other three muscles can readily be palpated on the surface. Support the extended leg at the ankle. The fusiform-shaped **rectus femoris** runs straight down the front of the thigh. Feel halfway down the lateral surface of the thigh for the **vastus lateralis.** The **vastus medialis** lies on the medial surface of the thigh just above the patella. Simultaneous contraction of these four muscles produces extension of the lower leg.
- Flexion of the lower leg is caused by simultaneous contraction of the three muscles of the **"hamstring" group**, the **semitendinosus, semimembranosus**, and the **biceps femoris.** All three arise mainly from the posterior inferior ischium and pass inferiorly on the back of the thigh. The bellies of the three muscles separate two-thirds the way down into the **semitendinosus** and **semimembranosus**, which pass medially, and the **biceps femoris**, which passes to the lateral side.
- Have the subject lie prone on a bench or table with the lower leg partially flexed. *With the subject's permission*, resist further flexing of the lower leg and palpate just distal to the ischial tuberosity. Feel for the contraction of the three hamstring muscles. Palpate the belly of the **biceps femoris** on the lateral side; palpate the **semitendinosus** on the medial side. The **semimembranosus** can best be felt distally on either side of the prominent tendon of the semitendinosus.

DORSAL AND PLANTAR FLEXION OF THE FOOT

- Ask the subject to dorsiflex the foot and feel for the medial tendon of the **tibialis anterior.** Follow it distally onto the foot to its insertion on the **medial cuneiform** and **base of the first metatarsal.** Follow the tendon upward as it transitions to a firm and narrow muscle lying lateral to the anterior border of the tibia.
- Have the subject stand. Feel the posterior surface of the heel where the broad **tendon of Achilles** is attached to the **calcaneus.** Follow the tendon upward, noting that it narrows and then broadly widens as you continue upward along the muscle fibers of the **gastrocnemius muscle.** Further upward, this muscle divides into the medial and lateral bellies, which extend to the **medial** and **lateral femoral epicondyles.** Feel for the separation between the two bellies. To palpate the **soleus muscle** that lies deep to the gastrocnemius, return to the Achilles tendon, relocate the transition to the gastrocnemius muscle, and extend your fingers to both sides to feel the outer edges of the soleus. The soleus primarily serves to stabilize the tibia during standing because it attaches to the tibia and fibula and not the femur. Have the subject gently raise the heel and feel the gastrocnemius contract. To feel the soleus contract, have the subject flex the knee and plantar flex the foot, which causes the gastrocnemius to relax and the soleus to contract.

Muscle Nerve Innervation

7

At the end of this chapter, you should be able to

1. Name the opening in the skull through which each of the cranial nerves pass.
2. Name the three major nerve plexuses in the body.
3. Name the innervation for the major arm muscles.
4. Describe the innervation for the diaphragm.
5. Describe the innervation for the major leg muscles.
6. Explain the "knee jerk" reflex.
7. Explain "foot drop."
8. Explain "driver's thigh."
9. List the three branches of the trigeminal nerve.
10. List the five branches of the facial nerve.

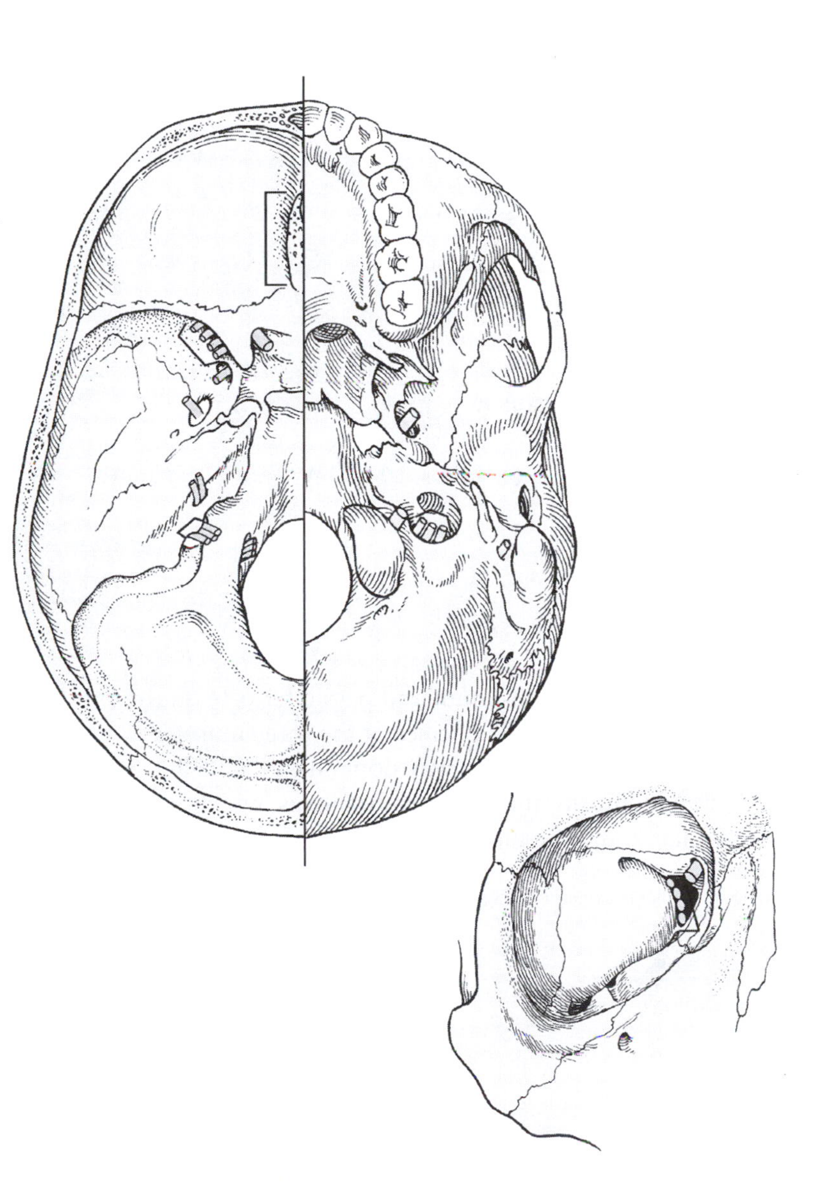

Word Search Puzzle

INSTRUCTIONS: Find and CIRCLE each of the listed terms in the Word Search Puzzle. (Terms may read from left to right, right to left, up, down, or diagonally.)

C	I	T	S	U	O	C	A	O	T	A	T	S	U	X	E	L	P
Z	P	R	A	K	L	U	M	B	O	S	A	C	R	A	L	Q	S
L	A	I	C	A	F	Y	X	T	C	B	S	Q	X	N	M	W	V
A	X	G	O	X	A	R	Z	U	C	R	I	B	I	F	O	R	M
R	P	E	U	I	C	O	D	R	A	L	U	B	I	F	W	X	Y
O	K	M	S	L	T	S	L	A	S	S	O	L	G	O	P	Y	H
M	X	I	T	L	O	S	B	T	F	S	U	T	A	E	M	X	Y
E	Z	N	I	A	R	E	R	O	T	O	M	O	L	U	C	O	C
F	U	A	C	R	Y	C	O	R	I	N	E	P	D	C	P	X	I
Z	Y	L	Y	Y	Q	C	V	K	B	L	D	M	W	T	F	Z	T
B	R	A	N	C	H	A	A	X	I	W	I	Y	I	Z	B	Q	A
X	S	U	G	A	V	V	L	C	A	B	A	C	X	Q	D	K	I
L	A	I	D	A	R	I	E	P	L	F	N	W	Q	Y	X	W	C
S	U	O	E	N	A	T	U	C	O	L	U	C	S	U	M	S	S

ACCESSORY
ACOUSTIC
AXILLARY
BRANCH
CRIBRIFORM
FACIAL
FEMORAL
FIBULAR
HYPOGLOSSAL
LUMBOSACRAL
MEATUS
MEDIAN
MUSCULOCUTANEOUS
OBTURATOR
OCULOMOTOR
OLFACTORY
OPTIC
OVALE
PLEXUS
RADIAL
SCIATIC
STATOACOUSTIC
TIBIAL
TRIGEMINAL
ULNAR
VAGUS

I. MATCHING

Group A: In the blank to the left of each cranial nerve, write the **LETTER(S)** of the passageway(s) in the skull through which that nerve passes. Branches of some cranial nerves pass though different openings; some passageways have more than one cranial nerve passing through them.

1. _____ Olfactory nerve
2. _____ Optic nerve
3. _____ Oculomotor nerve
4. _____ Trochlear nerve
5. _____ Trigeminal nerve
6. _____ Abducens nerve
7. _____ Facial nerve
8. _____ Statoacoustic nerve
9. _____ Glossopharyngeal nerve
10. _____ Vagus nerve
11. _____ Spinoaccessory nerve
12. _____ Hypoglossal nerve

A. Optic foramen
B. Hypoglossal canal
C. Foramen ovale
D. Internal acoustic meatus
E. Superior orbital fissure
F. Cribriform plate
G. Foramen rotundum
H. Jugular foramen
I. Stylomastoid foramen

Group B: In the blank to the left of each arm muscle, write the **LETTER(S)** of the nerves(s) that innervate that muscle. Each nerve may innervate more than one arm muscle and therefore may be used more than once.

1. _____ Biceps brachii
2. _____ Triceps brachii
3. _____ Supinator
4. _____ Coracobrachialis
5. _____ Brachioradialis
6. _____ Pronator teres
7. _____ Extensor carpi ulnaris
8. _____ Flexor carpi ulnaris

A. Ulnar nerve
B. Median nerve
C. Radial nerve
D. Musculocutaneous nerve

Group C: In the blank to the left of each leg muscle, write the **LETTER(S)** of the nerves(s) that innervate that muscle. Each nerve may innervate more than one leg muscle and therefore be used more than once.

1. _____ Adductor longus
2. _____ Quadriceps muscles
3. _____ Sartorius
4. _____ Hamstring muscles
5. _____ Fibularis longus
6. _____ Tibialis anterior
7. _____ Gastrocnemius
8. _____ Flexor digitorum longus

A. Femoral nerve
B. Tibial nerve
C. Superficial fibular nerve
D. Deep fibular nerve
E. Obturator nerve

III: SENTENCE COMPLETION

Circle the term in parentheses that best completes each statement.

1. The (**cervical, brachial**) plexus innervates the arm muscles.
2. Fibers from the (**cervical, brachial**) plexus combine with the spinoaccessory and hypoglossal cranial nerves to innervate some neck and pharyngeal muscles.
3. The (**spinoaccessory, hypoglossal, vagus, phrenic**) nerve innervates the diaphragm.
4. The flexors of the elbow are innervated by the (**axillary, musculocutaneous, median, ulnar**) nerve.
5. The deltoid and teres minor are innervated by the (**axillary, musculocutaneous, median, radial**) nerve.
6. The triceps brachii and anconeus are innervated by the (**axillary, musculocutaneous, median, radial**) nerve.
7. The leg adductor muscles and gracilis are innervated by the (**femoral, obturator, sciatic, common fibular**) nerve.
8. The (**tibial, fibular**) division of the sciatic nerve innervates the semimembranosus and semitendinosus muscles.
9. The sartorius and quadriceps femoris muscles are innervated by the (**sciatic, obturator, femoral**) nerve.
10. "Foot drop" is a condition caused by damage to the (**deep fibular, femoral, tibial**) nerve.
11. The (**femoral, fibular, tibial**) nerve innervates the gastrocnemius and soleus muscles.
12. Dilated pupils, wandering eye, and inability to visually follow the movements of a pencil indicate damage to the (**optic, oculomotor, trigeminal, facial**) nerve.
13. Loss of feeling in the face indicates damage to the (**glossopharyngeal, facial, trochlear, trigeminal**) nerve.
14. Loss of taste on the front of the tongue indicates damage to the (**glossopharyngeal, facial, hypoglossal, trigeminal**) nerve.

15. Damage to the (**ophthalmic, maxillary, mandibular**) division of the trigeminal nerve causes difficulty closing the jaw.
16. Bell's palsy is often due to inflammation and swelling of the (**glossopharyngeal, hypoglossal, facial, trigeminal**) nerve.
17. Difficulty swallowing may be caused by damage to the (**glossopharyngeal, hypoglossal**) nerve.
18. Loss of the sense of smell may be caused by damage to the (**oculomotor, trochlear, olfactory, abducens**) nerve.
19. All of the following nerves pass through the superior orbital fissure *except* the (**oculomotor, trochlear, optic, abducens**).
20. The (**glossopharyngeal, facial, statoacoustic, trigeminal**) nerve passes through the stylomastoid foramen.

IV. MULTIPLE CHOICE

In the blank to the left of each question, write the **LETTER** of the correct answer.

1. _____ The middle ear ossicle that attaches to the tympanic membrane is the _________.
   A. Hyoid
   B. Incus
   C. Malleus
   D. Stapes

2. _____ The phrenic nerve is considered part of the _________ plexus.
   A. Cervical
   B. Brachial
   C. Lumbosacral

3. _____ The sciatic nerve is consider part of the _________ plexus.
   A. Cervical
   B. Brachial
   C. Lumbosacral

4. _____ The biceps brachii muscle is innervated by the _________ nerve.
   A. Median
   B. Radial
   C. Axillary
   D. Musculocutaneous

5. _____ The supinator muscle is innervated by the _________ nerve.
   A. Median
   B. Radial
   C. Ulnar
   D. Musculocutaneous

6. _____ The deltoid muscle is innervated by the _________ nerve.
   A. Median
   B. Axillary
   C. Ulnar
   D. Radial

7. _____ The flexors of the hand are innervated by the _________ nerve.
   A. Median
   B. Ulnar
   C. Radial
   D. Musculocutaneous

8. _____ The only adductor of the thigh innervated by the femoral nerve is the _________.
   A. Adductor magnus
   B. Gracilis
   C. Pectineus
   D. Adductor longus

9. _____ The biceps femoris muscle is innervated by the _________ nerve.
   A. Obturator
   B. Femoral
   C. Pudendal
   D. Sciatic

10. _____ The branch of the sciatic nerve that innervates the gastrocnemius is the ________ nerve.

A. Tibial
B. Fibular

11. _____ The cranial nerve that innervates the muscles of mastication is the _______ nerve.

A. Oculomotor
B. Facial
C. Trigeminal
D. Hypoglossal

12. _____ The cranial nerve that first passes through the internal acoustic meatus and then the stylomastoid foramen is the _________ nerve.

A. Statoacoustic
B. Facial
C. Trigeminal
D. Glossopharyngeal

13. _____ The muscles of the dorsal surface of the foot are innervated by the ______ nerve.

A. Medial plantar
B. Fibular
C. Tibial
D. Lateral plantar

14. _____ Inability to straighten the leg at the knee might indicate damage to the ______nerve.

A. Fibular
B. Tibial
C. Femoral
D. Obturator

15. _____ Inability to evert the foot might suggest damage to the ________ nerve.

A. Fibular
B. Tibial
C. Femoral
D. Obturator

16. _____ Inability to dorsiflex the foot might indicate damage to the ______ nerve.

A. Deep fibular
B. Superficial fibular
C. Tibial
D. Lateral plantar

17. _____ Inability to extend the arm at the elbow might indicate damage to the ______ nerve.

A. Radial
B. Axillary
C. Ulnar
D. Musculocutaneous

18. _____ Crutches that are too long may cause "wrist drop" and the inability to extend the wrist and fingers due to damage of the ______ nerve.

A. Radial
B. Axillary
C. Ulnar
D. Musculocutaneous

19. _____ The branch of the facial nerve that innervates the upper jaw and lip is the _______.

A. Mandibular
B. Buccal
C. Zygomatic
D. Temporal

20. _____ The dermatomes of the upper anterior leg are innervated by the _______ nerve.

A. Sciatic
B. Femoral
C. Tibial
D. Obturator

V. FILL IN THE BLANK

Write the correct word in the blank spaces to complete each statement.

1. The plexus that innervates the buttocks and lower limbs is the ________________ plexus.
2. The ________________ nerve innervates the diaphragm.
3. The flexor muscles of the thigh are innervated by the ________________ nerve.
4. The ________________ nerve innervates the extensor muscles of the arm.
5. The pronators of the lower arm are innervated by the ________________ nerve.
6. The two nerves that innervate the flexors of the wrist and hand are the ________________ and the ________________ .
7. The obturator nerve innervates the ________________ muscles of the thigh.
8. The ________________ nerve is responsible for the patellar reflex.
9. The sciatic nerve is composed of ________________ and ________________ divisions.
10. The ________________ nerve is responsible for the sense of feeling on the face.
11. The ________________ nerve has both cranial and spinal components and innervates the ________________ and the ________________ muscles.
12. The muscles of facial expression are innervated by the ________________ nerve.
13. The ________________ nerve innervates the heart and many internal organs.
14. The ________________ nerve enters the skull through the internal acoustic meatus.
15. The inability to swallow could reflect damage to either one or both of the ________________ and ________________ nerves.
16. The skin on the head and neck is innervated by nerves from the ________________ region of the spinal cord.
17. The anterior surface of the leg is innervated by nerves from the ________________ region of the spinal cord.

18. The ________________ nerve innervates the muscles of the tongue.

19. The ________________ branch of the trigeminal nerve exits through the foramen ovale.

20. "Foot drop" is caused by damage to the ________________ nerve causing the ________________ muscle to be paralyzed.

VI. SHORT ANSWER

Answer each of the following questions in the space provided.

1. A patient has damage to cervical nerves 3 and 4. Specifically, how would that affect breathing and why?

2. A patient has damage to the axillary branch of the brachial plexus. What muscular deficits would the patient exhibit?

3. What symptoms would you expect in a patient with compression of the sciatic nerve?

4. Explain what happens when the patellar ligament is tapped with a reflex hammer. Which muscles respond? Which nerve is responsible for the action?

VII. CASE STUDIES

A. One morning a woman woke up and found that when she went to brush her teeth, her lips were paralyzed on one side, and her face and eyelid were drooping on the same side. She had trouble eating because food dribbled out of her mouth. She went to the emergency room thinking she had had a stroke. Examination revealed this was not the case.

1. What is the most likely diagnosis?

2. Which cranial nerve is affected?

3. Based on the symptoms, which branches of this nerve were involved?

4. Which muscles were affected to cause the various symptoms?

B. A long-distance truck driver went to a chiropractor complaining of pain in the buttock and radiating down the outer thigh. Examination revealed that there was no disk injury.

1. Name one probable diagnosis for this condition.

2. Name the painful buttock muscles.

3. Name the painful thigh muscles.

4. Which major nerve was being compressed?

5. Name a similar syndrome caused by the same nerve passing through a hip muscle.

Appendix A

General Answer Key

CHAPTER 1
Skeleton and Fractures

I. ILLUSTRATION IDENTIFICATION

Views of skeleton

1. Skull
   a. Cranium
   b. Facial bones
   c. Mandible
2. Vertebral column
3. Sternum
4. Ribs
5. Clavicle
6. Scapula
7. Humerus
8. Radius
9. Ulna
10. Carpals
11. Metacarpals
12. Phalanges
13. Os coxa
14. Femur
15. Patella
16. Tibia
17. Fibula
18. Tarsals
19. Metatarsals
20. Phalanges

Lateral and anterior skull

1. Frontal bone
2. Temporal bone
3. Parietal bone
4. Occipital bone
5. Nasal bone
6. Maxilla
7. Mandible
8. Sphenoid bone
9. Lacrimal bone
10. Zygomatic bone
11. Mastoid process
12. External auditory meatus
13. External occipital protuberance
14. Coronal suture
15. Squamosal suture
16. Lambdoidal suture
17. Zygomatic process
18. Styloid process
19. Sagittal suture
20. Bony orbit
21. Mental foramen
22. Supraorbital foramen
23. Nasal septum
24. Infraorbital foramen
25. Mandibular condyle
26. Maxillary alveolar process

Superior and inferior skull

1. Frontal bone
2. Parietal bone
3. Occipital bone
4. Zygomatic arch
5. Coronal suture
6. Sagittal suture
7. Lambdoidal suture
8. Palatine process of maxilla
9. Incisive foramen
10. Palatine bone
11. Vomer bone
12. Sphenoid bone
13. Zygomatic bone
14. Zygomatic process
15. Styloid process
16. Mastoid process
17. Occipital condyle
18. Occipital protuberance
19. Superior nuchal line
20. Inferior nuchal line
21. Foramen ovale
22. Foramen spinosum
23. Foramen lacerum
24. Carotid canal
25. Jugular foramen
26. Foramen magnum
27. Stylomastoid foramen
28. Basilar process of occipital
29. Jugular process

Internal skull

1. Frontal bone
2. Crista galli
3. Lesser wing of sphenoid
4. Greater wing of sphenoid
5. Temporal bone
6. Petrous part of temporal
7. Parietal bone
8. Occipital bone
9. Anterior cranial fossa
10. Middle cranial fossa
11. Posterior cranial fossa
12. Sella turcica
13. Cribriform plate
14. Optic foramen
15. Foramen rotundum
16. Foramen ovale
17. Foramen spinosum
18. Foramen lacerum

19. Internal acoustic meatus
20. Jugular foramen
21. Hypoglossal canal
22. Foramen magnum

Mandible

1. Oblique line
2. Mandibular foramen
3. Mandibular symphysis
4. Coronoid process
5. Mandibular angle
6. Mandibular condyle
7. Mandibular arch
8. Lateral fossa
9. Ramus of mandible
10. Body of mandible
11. Alveolar margin
12. Incisive fossa
13. Inferior mental spine

Cervical vertebrae and hyoid

1. Mandible
2. Mastoid process
3. Styloid process
4. Cervical vertebra
5. Hyoid bone
6. Thyroid cartilage
7. Body of hyoid
8. Lesser cornu
9. Greater cornu

Sternum and thoracic cage

1. First thoracic vertebra
2. Twelfth thoracic vertebra
3. Manubrium of sternum
4. Body of sternum
5. Xiphoid process
6. Costal cartilage
7. Jugular notch
8. Sternal angle
9. Floating rib
10. Costal notches

Ribs

1. Head
2. Articular facets
3. Neck
4. Tubercle
5. Shaft
6. Costal angle
7. Scalene tubercle
8. Superior border
9. Costal groove
10. Inferior border

Vertebral column, cervical

1. Vertebral foramen
2. Transverse process
3. Lamina
4. Superior articulating facet
5. Spinous process
6. Body
7. Transverse foramen
8. Pedicle
9. Anterior tubercle
10. Posterior tubercle
11. Anterior arch
12. Posterior arch
13. Odontoid process
14. Anterior tubercle of transverse process
15. Posterior tubercle of transverse process

Vertebral column, lower

1. Vertebral foramen
2. Transverse process
3. Lamina
4. Superior articulating facet
5. Spinous process
6. Body
7. Pedicle
8. Facet for tubercle of rib
9. Demifacet for head of rib
10. Sacral promontory
11. Transverse lines
12. Anterior sacral foramen
13. Posterior sacral foramen
14. Sacral hiatus
15. Medial sacral crest
16. Sacral canal
17. Sacroiliac articulating surface
18. Coccyx
19. Ala of sacrum
20. Lateral sacral crest
21. Mammillary process

Clavicle

1. Acromial extremity
2. Sternal extremity
3. Conoid tubercle
4. Body of clavicle
5. Costal tuberosity

Scapula

1. Vertebral border
2. Axillary border
3. Acromion process
4. Coracoid process
5. Inferior angle
6. Superior angle
7. Glenoid fossa
8. Infraglenoid tubercle
9. Supraglenoid tubercle
10. Spine
11. Infraspinous fossa
12. Supraspinous fossa
13. Subscapular fossa
14. Scapular notch

Humerus

1. Head
2. Anatomical neck
3. Surgical neck
4. Greater tubercle
5. Lesser tubercle
6. Intertubercular groove

7. Deltoid tuberosity
8. Radial groove
9. Lateral supracondylar ridge
10. Lateral epicondyle
11. Medial epicondyle
12. Medial supracondylar ridge
13. Radial fossa
14. Capitulum
15. Coronoid fossa
16. Trochlea
17. Olecranon fossa

Radius and ulna

1. Head
2. Neck
3. Radial tuberosity
4. Anterior oblique line
5. Posterior oblique line
6. Interosseous border
7. Styloid process
8. Dorsal tubercle
9. Olecranon process
10. Trochlear notch
11. Coronoid process
12. Radial notch
13. Ulnar tuberosity
14. Interosseous border
15. Styloid process
16. Head
17. Supinator crest

Hand

Carpals

1. Scaphoid
2. Lunate
3. Triquetrum
4. Pisiform
5. Trapezium
6. Trapezoid
7. Capitate
8. Hamate

Metacarpals

9. Base
10. Shaft
11. Head

Phalanges

12. Proximal
13. Middle
14. Distal

Os coxa

Ilium

1. Iliac crest
2. Iliac fossa
3. Posterior gluteal line
4. Anterior gluteal line
5. Inferior gluteal line
6. Anterior superior iliac spine
7. Anterior inferior iliac spine
8. Posterior superior iliac spine
9. Posterior inferior iliac spine
10. Greater sciatic notch

Ischium

11. Ischial tuberosity
12. Ischial spine
13. Lesser sciatic notch
14. Ramus of ischium

Pubis

15. Pubic Crest
16. Pubic symphysis
17. Superior ramus of pubis
18. Inferior ramus of pubis
19. Acetabulum
20. Obturator foramen

Femur

1. Head
2. Neck
3. Greater trochanter
4. Lesser trochanter
5. Intertrochantic line
6. Intertrochantic crest
7. Trochanteric fossa
8. Quadrate tubercle
9. Gluteal tuberosity
10. Linea aspera
11. Adductor tubercle
12. Medial epicondyle
13. Lateral epicondyle
14. Patellar surface
15. Medial supracondylar line
16. Lateral supracondylar line
17. Medial condyle
18. Lateral condyle
19. Popliteal surface
20. Intercondylar notch

Tibia and fibula

Fibula

1. Apex of head
2. Head
3. Neck
4. Interosseous border
5. Lateral malleolus
6. Shaft

Tibia

7. Medial condyle
8. Lateral condyle
9. Tibial tuberosity
10. Soleal line
11. Shaft
12. Interosseous border
13. Medial malleolus

Foot

1. Talus
2. Calcaneus
3. Navicular
4. Cuboid
5. First cuneiform
6. Second cuneiform
7. Third cuneiform
8. Metatarsals
9. Proximal phalanges
10. Medial phalanges
11. Distal phalanges

II. MATCHING

Group A

1. B
2. F
3. J
4. A
5. G
6. C
7. B
8. E
9. I
10. D

Group B

1. A, B, E, G, I, L
2. A, E, F, G, H, K
3. B, C, D, I, K
4. E, F, G, H, M
5. D, E
6. E, F, J

III. SENTENCE COMPLETION

1. Thoracic
2. Epicondyles
3. Comminuted
4. Talus
5. Foramen
6. Epiphyses
7. Sever's disease
8. Cervical
9. Glenoid fossa
10. Radius swivels across the ulna
11. Clavicles and scapulae
12. 7
13. Magnum
14. Maxilla
15. Xiphoid process
16. Atlas
17. Trapezium
18. Navicular
19. Avulsion
20. Hip

IV. MULTIPLE CHOICE

1. C
2. D
3. D
4. D
5. B
6. D
7. A
8. B
9. B
10. B
11. C
12. B
13. B
14. E
15. C

V. FILL IN THE BLANK

1. Scapula, clavicle, and humerus
2. Sagittal
3. Greenstick fracture
4. Compound
5. Calcaneus
6. Cheek bone
7. Patella
8. Olecranon process
9. Clavicle
10. Mandible and dentary
11. Breast bone
12. Pollux
13. Hip bone
14. Shoulder blade
15. Big toe

CHAPTER 2
Articulations and Body Motions

I. BODY MOTIONS IDENTIFICATION

■ Views of skeleton

Figure 2.1
A. Abduction
B. Adduction

Figure 2.2
A. Flexion
B. Extension

Figure 2.3
A. Pronation
B. Supination

Figure 2.4
A. Lateral Rotation
B. Medial Rotation

Figure 2.5
A. Circumduction

Figure 2.6
A. Plantar flexion
B. Dorsiflexion

Figure 2.7
A. Inversion
B. Eversion

Figure 2.8
A. Elevation
B. Depression

II. MATCHING

Group A
1. F
2. E
3. B
4. A
5. D

Group B
1. A, E, F
2. A, B, D, F
3. A
4. F
5. A, B, F
6. E

Group C
1. I
2. H
3. A
4. C
5. E
6. D
7. B
8. F
9. J
10. G

III. SENTENCE COMPLETION

1. Symphysis
2. Suture
3. Condyloid
4. Gliding
5. Annular
6. Gliding
7. Pivot
8. Symphysis
9. Hinge
10. Ball and socket
11. Radial collateral
12. Iliofemoral
13. Adduction
14. Supination
15. Plantar flexion

IV. MULTIPLE CHOICE

1. B
2. B
3. A
4. B
5. D
6. C
7. D
8. C
9. C
10. D
11. C
12. C
13. C
14. B
15. D

V. FILL IN THE BLANK

1. Suture, syndesmosis, gomphosis
2. Synchondrosis
3. Synovial
4. Condyloid
5. Carpal, metacarpal
6. Pivot
7. Plantar flexion
8. Protraction
9. Anterior cruciate, posterior cruciate
10. Ligamentum teres
11. Long plantar, short plantar, deltoid, spring
12. Plantar aponeurosis
13. Rotation
14. Flexion
15. Abduction

CHAPTER 3

Muscles of the Head, Neck, and Torso

I. ILLUSTRATION IDENTIFICATION

(Answers are on the reverse side of illustration pages.)

II. MATCHING

Group A

1. J
2. E
3. D
4. G
5. A
6. H
7. B
8. F
9. I
10. C

Group B

1. F
2. E
3. H
4. A
5. G
6. B
7. D
8. C

Group C

1. D
2. I
3. C
4. J
5. F
6. A
7. H
8. G
9. B
10. E

III. SENTENCE COMPLETION

1. Temporoparietalis
2. Orbicularis oculi
3. Corrugator supercilii
4. VII
5. Levator labii superioris
6. Buccinator
7. Masseter
8. Platysma
9. Digastric
10. Sternohyoid
11. Stylohyoid
12. Obliquus capitis superior
13. Scalenes medius
14. Multifidus
15. Intertransversarii
16. Quadratus lumborum
17. Rectus femoris
18. Serratus posterior inferior
19. Phrenic
20. External intercostals

IV. MULTIPLE CHOICE

1. A
2. C
3. D
4. C
5. B
6. C
7. D
8. D
9. C
10. B
11. C
12. A
13. D
14. A
15. D
16. C
17. C
18. A
19. B
20. C

V. FILL IN THE BLANK

1. Occipitalis and frontalis
2. Levator palpebrae superioris
3. Procerus
4. Nasalis
5. Masseter and temporalis
6. Medial and lateral pterygoideus
7. Glosso
8. Styloid process, hyoid, digastric
9. Sternohyoid, thyrohyoid
10. Sternocleidomastoid, sternum; clavicle, mastoid
11. Rectus capitis anterior
12. Rectus capitis posterior major and minor
13. Iliocostalis, longissimus, spinalis
14. Levator costarum
15. Internal intercostals, subcostals, transversus thoracis

CHAPTER 4

Muscles of Shoulder, Arm, and Hand

I. ILLUSTRATION IDENTIFICATION

(Answers are on the reverse side of illustration pages.)

II. MATCHING

Group A

1. C
2. E
3. J
4. F
5. H
6. I
7. B
8. A
9. D
10. G

Group B

1. E
2. H
3. C
4. G
5. B
6. F
7. A
8. D

Group C

1. C
2. F
3. G
4. A
5. H
6. E
7. B
8. D
9. J
10. I

III. SENTENCE COMPLETION

1. Abduction of arm
2. Subclavius
3. Levator scapulae
4. Serratus anterior
5. Latissimus dorsi
6. Posterior
7. Teres major
8. Adductor pollicis
9. Teres minor
10. Long head
11. Anconeus
12. Ulnar
13. Flexor pollicis longus
14. Flexor digitorum superficialis
15. Pronator quadratus
16. Extensor pollicis longus
17. Abductor pollicis brevis
18. Supraspinatus
19. Palmaris brevis
20. Extensor carpi ulnaris

IV. MULTIPLE CHOICE

1. C
2. A
3. B
4. C
5. B
6. A
7. D
8. B
9. A
10. A
11. C
12. D
13. C
14. A
15. C
16. B
17. C
18. C
19. A
20. C

V. FILL IN THE BLANKS

1. Subclavius
2. Triceps brachii
3. Flexor carpi radialis
4. Pectoralis minor
5. Latissimus dorsi
6. Rhomboideus major
7. Latissimus dorsi
8. Deltoid
9. Supraspinatus, infraspinatus
10. Flexion of elbow
11. Brachioradialis
12. Supinator
13. Palmaris longus
14. Flexor digitorum superficialis
15. Flexor pollicis longus
16. Abductor pollicis
17. Extensor digiti minimi
18. Opponens pollicis
19. Lumbricales
20. Interossei

CHAPTER 5

Muscles of the Hip, Thigh, and Lower Leg

I. ILLUSTRATION IDENTIFICATION

(Answers are on the reverse side of illustration pages.)

II. MATCHING

Group A

1. C
2. E
3. A
4. D
5. B

Group B

1. E
2. G
3. F
4. B
5. H
6. A
7. D
8. C

Group C

1. E
2. C
3. H
4. B
5. A
6. G
7. D
8. F

III. SENTENCE COMPLETION

1. Gluteus maximus
2. Gluteus minimus
3. Tensor fascia latae
4. Sartorius
5. Biceps femoris
6. Rectus femoris
7. Quadriceps muscles
8. Biceps femoris
9. Sartorius
10. Adductors
11. Gluteus maximus
12. Gastrocnemius
13. Soleus
14. Plantaris
15. Popliteus
16. Flexor digitorum longus
17. Tibialis anterior
18. Tibialis anterior
19. Lumbricales
20. Adductor hallucis

IV. MULTIPLE CHOICE

1. D
2. C
3. D
4. D
5. B
6. B
7. C
8. C
9. C
10. D
11. A
12. A
13. B
14. D
15. A
16. B
17. B
18. C
19. B
20. C

V. FILL IN THE BLANK

1. Flexor digiti minimi
2. Adductor hallucis
3. Quadratus plantae or flexor digitorum longus
4. Extensor digitorum longus and fibularis tertius
5. Abductor hallucis
6. Tibialis anterior
7. Flexor hallucis longus and extensor hallucis
8. Calcaneus
9. Gastrocnemius
10. Adductor magnus
11. Pectineus
12. Adductor
13. Biceps femoris, semitendinosus, semimembranosus
14. Vastus lateralis
15. Hematoma
16. Rectus femoris, vastus lateralis, v. medialis, v. intermedius
17. Iliacus, sartorius, rectus femoris, or tensor fascia latae
18. Gluteus medius and minimus
19. Laterally
20. Piriformis

CHAPTER 6

Functional Muscle Groups

I. ILLUSTRATION IDENTIFICATION

Elevators of scapula

1. Levator scapula
2. Trapezius (upper)
3. Rhomboideus minor
4. Rhomboideus major

Depressors of scapula

1. Pectoralis minor
2. Serratus anterior
3. Pectoralis major
4. Trapezius (lower)

Protractors of scapula

1. Pectoralis minor
2. Serratus anterior

Retractors of scapula

1. Rhomboideus minor
2. Trapezius (middle)
3. Rhomboideus major

Upward rotators of scapula

1. Trapezius (upper)
2. Trapezius (lower)
3. Serratus anterior

Downward rotators of scapula

1. Rhomboideus minor
2. Rhomboideus major
3. Levator scapula
4. Pectoralis minor

Medial rotators of humerus

1. Deltoid (anterior)
2. Pectoralis major
3. Subscapularis
4. Teres major
5. Latissimus dorsi

Lateral rotators of humerus

1. Infraspinatus
2. Deltoid (posterior)
3. Teres minor

Flexors of humerus

1. Deltoid (anterior)
2. Pectoralis major (clavicular head)
3. Coracobrachialis
4. Biceps brachii

Extensors of humerus

1. Pectoralis major (sternal head)
2. Deltoid (posterior)
3. Triceps brachii (long head)
4. Teres major
5. Latissimus dorsi

Abductors of humerus

1. Supraspinatus
2. Deltoid (middle and posterior)
3. Infraspinatus

Adductors of humerus

1. Deltoid (anterior)
2. Coracobrachialis
3. Pectoralis major
4. Teres major
5. Latissimus dorsi

Flexors of elbow

1. Biceps brachii
2. Brachialis
3. Pronator teres
4. Brachioradialis

Extensors of elbow

1. Triceps brachii
2. Anconeus

Supinators of forearm

1. Biceps brachii
2. Supinator
3. Brachioradialis

Pronators of forearm

1. Pronator teres
2. Brachioradialis
3. Pronator quadratus

Flexors of wrist

1. Flexor carpi radialis
2. Palmaris longus
3. Flexor carpi ulnaris
4. Flexor digitorum superficialis
5. Flexor digitorum profundus

Extensors of wrist

1. Extensor carpi radialis longus
2. Extensor carpi radialis brevis
3. Extensor carpi ulnaris
4. Extensor digitorum

Adductors of wrist

1. Extensor carpi ulnaris
2. Flexor carpi ulnaris

Abductors of wrist

1. Extensor carpi radialis longus
2. Extensor carpi radialis brevis
3. Abductor pollicis longus
4. Flexor carpi radialis
5. Flexor pollicis longus

Abductors of thumb and digits

1. Dorsal interossei
2. Abductor digiti minimi
3. Abductor pollicis brevis
4. Abductor pollicis longus

Adductors of thumb and digits

1. Adductor pollicis
2. Palmar interossei

Extension of thumb and digits

1. Extensor pollicis longus tendon
2. Extensor pollicis brevis tendon
3. Dorsal interossei
4. Extensor digitorum tendon
5. Extensor digiti minimi tendon

Flexion—opposition of thumb and digits

1. Flexor pollicis longus tendon
2. Flexor pollicis brevis
3. Opponens pollicis
4. Flexor digitorum profundus tendon
5. Flexor digitorum superficialis tendon
6. Lumbricales
7. Opponens digiti minimi

Lateral rotators of hip

1. Gluteus maximus
2. Gluteus medius
3. Gluteus minimus
4. Piriformis
5. Gemellus superior
6. Quadratus femoris
7. Gemellus inferior
8. Obturator externus

Medial rotators of hip

1. Gluteus medius
2. Gluteus minimus
3. Tensor fascia latae

Extensors of hip

1. Gluteus medius
2. Gluteus maximus
3. Biceps femoris
4. Semitendinosus
5. Adductor magnus
6. Semimembranosus

Flexors of hip

1. Psoas major
2. Iliacus
3. Tensor fascia latae
4. Rectus femoris
5. Pectineus
6. Adductor brevis
7. Adductor longus
8. Adductor magnus
9. Sartorius
10. Gracilis

Abductors of hip

1. Gluteus medius
2. Gluteus minimus
3. Gluteus maximus
4. Tensor fascia latae
5. Sartorius

Adductors of hip

1. Pectineus
2. Adductor brevis
3. Adductor longus
4. Gracilis
5. Adductor magnus

Lateral rotators of knee

1. Biceps femoris

Medial rotators of knee

1. Gracilis
2. Semitendinosus
3. Semimembranosus
4. Sartorius
5. Popliteus

Extensors of knee

1. Tensor fascia latae
2. Vastus intermedius
3. Vastus lateralis
4. Rectus femoris tendon
5. Vastus medialis

Flexors of knee

1. Gracilis
2. Semitendinosus
3. Semimembranosus
4. Gastrocnemius
5. Sartorius
6. Biceps femoris
7. Plantaris

Dorsiflexors of ankle

1. Extensor digitorum longus
2. Fibularis tertius
3. Tibialis anterior
4. Extensor hallucis longus

Plantar flexors of ankle

1. Plantaris
2. Gastrocnemius
3. Flexor hallucis longus
4. Fibularis longus
5. Fibularis brevis
6. Flexor digitorum longus
7. Soleus

Invertors of foot

1. Tibialis posterior
2. Flexor digitorum longus
3. Flexor hallucis longus

Evertors of foot

1. Fibularis longus
2. Fibularis brevis
3. Extensor digitorum longus
4. Fibularis tertius

II. MATCHING

Group A

1. B, E, F, G
2. A, B, C, D
3. H, I
4. A, D
5. E, F, G
6. E, F, G

Group B

1. D, E, F, G
2. G, H
3. C, F, G
4. C, D, E
5. A, B, C, D
6. B, C,

III. COMPLETION

1. Latissimus dorsi
2. Levator scapula
3. Posterior deltoid
4. Brachialis
5. Teres minor
6. Brachialis
7. Extensor carpi ulnaris
8. Sartorius
9. Semitendinosus
10. Biceps femoris
11. Tibialis anterior
12. Fibularis longus
13. Extensor hallucis longus
14. Gastrocnemius
15. Sartorius
16. Pubis
17. Ilium
18. Metacarpals
19. Olecranon process
20. Scapula

IV. MULTIPLE CHOICE

1. C
2. D
3. B
4. A
5. C
6. C
7. B
8. D
9. D
10. B
11. C
12. A
13. B
14. B
15. C
16. A
17. A
18. C
19. C
20. C

V. FILL IN THE BLANK

1. Tibialis anterior
2. Tensor fascia latae
3. Biceps femoris, semimembranosus, semitendinosus
4. Biceps femoris
5. Sartorius
6. Gluteus minimus
7. Opponens pollicis
8. Extensor carpi ulnaris, flexor carpi ulnaris
9. Flexors of the wrist
10. Anconeus
11. Quadriceps
12. Pectoralis minor, serratus anterior
13. Medial rotators of humerus
14. Downward rotators of scapula, retractors of scapula
15. Biceps brachii, brachialis, brachioradialis, pronator teres
16. Popliteus and plantaris
17. Palmar interossei
18. Dorsal
19. Flexor digitorum superficialis and flexor digitorum profundus
20. Pronator quadratus

CHAPTER 7

Muscle Nerve Innervation

I. MATCHING

Group A

1. F
2. A
3. E
4. E
5. E, C, G
6. E
7. D, I
8. D
9. H
10. H
11. H
12. B

Group B

1. D
2. C
3. C
4. D
5. C
6. B
7. C
8. A

Group C

1. E
2. A
3. A
4. B
5. C
6. D
7. B
8. B

II. COMPLETION

1. Brachial
2. Cervical
3. Phrenic
4. Musculocutaneous
5. Axillary
6. Radial
7. Obturator
8. Tibial
9. Femoral
10. Deep fibular
11. Tibial
12. Oculomotor
13. Trigeminal
14. Facial
15. Mandibular
16. Facial
17. Glossopharyngeal
18. Olfactory
19. Optic
20. Facial

III. MULTIPLE CHOICE

1. C
2. A
3. C
4. D
5. B
6. B
7. A
8. C
9. D
10. A
11. C
12. B
13. B
14. C
15. A
16. A
17. A
18. A
19. B
20. B

IV. FILL IN THE BLANK

1. Lumbosacral
2. Phrenic
3. Sciatic
4. Radial
5. Median
6. Median and ulnar
7. Adductor
8. Femoral
9. Tibial and fibular
10. Trigeminal
11. Spinoaccessory, sternocleidomastoid, trapezius
12. Facial
13. Vagus
14. Statoacoustic
15. Vagus and glossopharyngeal
16. Cervical
17. Lumbar
18. Hypoglossal
19. Mandibular
20. Deep fibular, tibialis anterior

Short Answer and Case Study Answer Keys

Appendix

B

CHAPTER 1

Skeleton and Fractures

VI. SHORT ANSWER

1. A. The costal cartilages of the seven "true" ribs (1–7) articulate directly and individually with the sternum.

   B. The costal cartilages of the three "false" ribs (8–10) attach to one another and then to cartilage of the 7th pair; thus they articulate indirectly with the sternum.

   C. The two floating ribs (11 and 12) do not articulate with the sternum.

   Note: The term "false" ribs is frequently used to refer to ribs 8–12.

2. The functional advantage is that the fused bones provide protection for the underlying soft tissues.

3. A. Humerus

   B. Vertebrae

   C. Occipital

   D. Fracture

   E. Radial

   F. Thoracic

4. A displaced fracture is one in which the ends of the bones are no longer in correct anatomical alignment. Three examples are spiral, avulsion, and compound fractures.

5. An apophysis is an outgrowth of a bone for attachment for tendons and ligaments. An epiphysis is the end of a long bone usually including one or more articulating surfaces.

VII. COMPARISON OF MALE AND FEMALE PELVISES

A

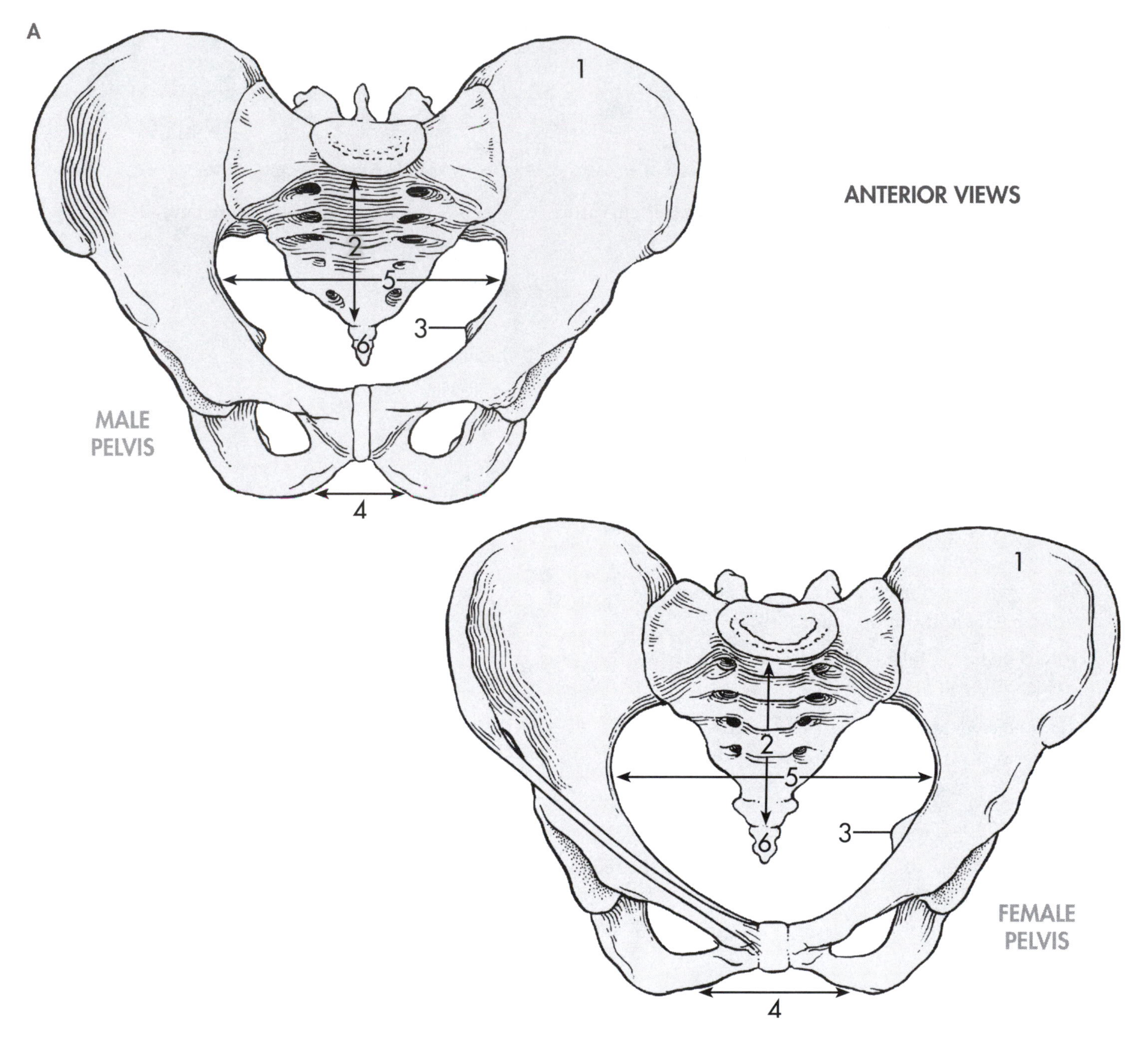

B

LANDMARK	FEMALE	MALE
1. Iliac Crest	Flares more laterally	Higher and less lateral flare
2. Sacrum	Shorter, wider, flatter curvature	Longer, narrower, more curved
3. Ischial Spine	Short, more posterior protrusion	Longer, more inward protrusion
4. Pubic Angle	Wide, concave shape, greater than 90 degrees	Narrow, V-shaped, 90 degrees or less
5. Pelvic Inlet	Wide and oval-roundish	"Heart shaped"
6. Coccyx	Less ventro-anterior curvature	Greater ventro-anterior curvature

C The combination of anatomical differences in the female pelvis compared to the male pelvis provide a larger inlet and outlet for birth.

VIII. CASE STUDIES

A
1. A broken bone
2. The clavicle
3. Sternocleidomastoid
4. The lateral half of the clavicle that articulates with the acromion process would have no anterior support, causing the shoulder to sag.

B
1. The fibula
2. In a spiral fracture, the distal end of the bone is twisted so that it no longer is in correct anatomical alignment.
3. The muscles that overlie the fibula are the fibularis longus and brevis, the extensor digitorum longus, and the extensor hallucis longus
4. Fibular nerve
5. Dorsiflexion of the ankle, extension of the toes, and eversion of the foot

CHAPTER 2

Articulations and Body Motions

VI. SHORT ANSWER

1. The shoulder joint has a greater range of motion and is less stable than the hip joint because of the shallow curvature of the glenoid fossa versus the much deeper acetabular fossa.
2. The temporomandibular joint allows for movement in two directions. The temporalis and masseter muscles elevate the jaw, and the digastric muscles lower it; the medial and lateral pterygoids also move it side-to-side.
3. The menisci are fibrocartilage pads that lie between the ends of the femur and tibia. They act as shock absorbers and minimize side-to-side movement of the femur on the tibia.
4. A bursa is a synovial-fluid-filled sac that cushions tendons and ligaments at friction points around a joint.
5. The malleoli stabilize the joint and serve as points of attachment for muscles and ligaments.

VII. CASE STUDIES

A

1. The anterior cruciate ligament and the lateral collateral ligament
2. The menisci and the articular cartilage
3. The cruciate ligament provides for front-to-back stability and prevents hyperextension of the knee.
4. The menisci act as shock absorbers and help stabilize the joint, preventing excessive side-to-side motion.

B

1. The bodies of adjacent lumbar vertebrae articulate by fibrocartilaginous intervertebral discs, and the superior and inferior facets of the transverse processes articulate by gliding synovial joints.
2. An intervertebral disc could become dislocated (protruded) or compressed and inflamed; the articular cartilage of adjacent facets may become thin, resulting in abrasion of surfaces and development of an arthritic condition, osteoarthritis.
3. A dislocated or compressed intervertebral disc would press on the nerve passing through the adjacent lateral foramen and produce pain in the muscles innervated by that nerve.
4. The patient most likely would be diagnosed with facet syndrome of L4–L5 with concurrent spasm of the paraspinal muscles to minimize movement of the effected vertebrae. The patient indicated that he had pain trying to straighten his back, which would have exerted more pressure on the facets. The doctor noted the pain in the spinal muscles adjacent to the vertebrae and noticed the rigid body posture in this region. An intervertebral disc pressure on a major spinal nerve at this level would have been expected to produce more extensive chronic pain, including on the lower leg.

CHAPTER 3

Muscles of the Head, Neck, and Torso

VI. SHORT ANSWER

1. A person with Bell's palsy would likely have a drooping eyelid, drooping facial muscles, an uneven smile, and possible drooling, all on one side of the face.
2. During anesthesia, the pharyngeal muscles and uvula collapse, blocking he airway and jeopardizing breathing.
3. The sternocleidomastoid, rectus capitis anterior, long capitis, and longus colli all flex the neck.
4. Back strain is caused by some degree of stretching and tearing of muscle fibers or ligaments in the lumbar region due to excessive extension or rotation of the vertebral column without proper back support. The erector spinae groups are the primary muscles affected.
5. The erector spinae are the primary muscle antagonists to the rectus abdominis muscle.

VII. CASE STUDIES

A

1. Whiplash injury
2. The brachial plexus
3. The posterior extensors of the neck: the rectus posterior muscles and the obliquus capitis muscles
4. Musculocutaneous nerve

B

1. The lower portions of the erector spinae muscles, the psoas major and iliacus
2. The sciatic nerve
3. The hamstrings, adductor magnus, tibialis anterior, extensor muscles of the lower leg, gastrocnemius, soleus, flexor digitorum longus, and tibialis posterior
4. The tibial branch of the sciatic nerve innervates the gastrocnemius and soleus; the fibular division innervates the tibialis anterior and the other muscles involved in dorsiflexion. With both dorsiflexion and plantar flexion being impeded, the individual will limp.

CHAPTER 4

Muscles of the Arm and Shoulder

VI. SHORT ANSWER

1. The referred pain pattern for the sternalis muscles is similar to the pain of a myocardial infarction or angina pectoralis and therefore may lead to a false diagnosis.
2. The levator scapularis, upper trapezium, and the rhomboideus major and minor muscles work together to raise the scapula.
3. "Golfer's elbow" results from repetitive use of the superficial muscles of the forearm, straining the common flexor tendon. The flexor carpi radialis is the most commonly affected muscle.
4. Carpal tunnel syndrome is caused by inflammation of the flexor retinaculum, decreasing the size of the carpal tunnel and compressing the median nerve.
5. The three major actions of these muscles are pronation of the lower arm, flexion of the wrist, and flexion of the digits.

VII. CASE STUDIES

A

1. The lateral end of the clavicle and the acromion process of the scapula
2. The supraspinatus muscle
3. The acromion process of the scapula
4. Abduction of the humerus
5. Subscapularis, teres major, and coracobrachialis

B

1. Carpal tunnel syndrome
2. Flexor digitorum superficialis and profundus, flexor carpi radialis, and flexor pollicis longus
3. Opponens pollicis, flexor pollicis brevis, abductor pollicis
4. Flexion, abduction, and opposition
5. Median and ulnar nerves

CHAPTER 5

Muscles of the Hip, Thigh, and Lower Leg

VI. SHORT ANSWER

1. The muscles most likely involved in "groin strain" are the psoas major, the iliacus, and the adductor muscles.
2. Chondromalacia patella, or "runner's knee," is a softening of the cartilage around the patella caused by compression of the knee, quadriceps imbalance, or a blow to the patella.
3. The symptoms of a ruptured Achilles tendon are a sharp pain, a bulging calf muscle, and an inability to plantar flex the ankle. The gastrocnemius and soleus muscles would be impaired.
4. "Foot drop" is a condition in which the foot must be lifted high to prevent it from dragging. It is caused by paralysis of the tibialis anterior muscle.
5. The three compartments of the lower leg are the anterior, posterior, and lateral compartments. The primary actions of the anterior compartment muscles are dorsiflexion of the ankle and extension of the toes; the primary actions of the posterior compartment are plantar flexion of the ankle and flexion of the toes; and the primary actions of the lateral or fibular compartment are plantar flexion and eversion of the foot.

VII. CASE STUDIES

A

1. Adductor group
2. Either the pectineus or the adductor brevis
3. The tendon being separated or partially separated from the point of attachment on the femur
4. Hematoma or inguinal hernia

B

1. "Shin splints"
2. Tibialis anterior
3. The deep fibular branch of the fibular nerve
4. "Foot drop"

CHAPTER 6
Functional Muscle Groups

VI. SHORT ANSWER

1 A. Flexor digitorum: muscle bends fingers and toes

B. Temporalis: positioned over the temporal bone

C. Biceps brachii: muscle with two heads; located on arm

D. Deltoid: muscle shape is like a triangle or the Greek letter "delta"

E. External oblique: muscle with outer layer of fibers running diagonally

F. Adductor magnus: muscle that is the largest of the leg adductors

G Coracobrachialis: muscle located on arm with origin from coracoid process

2 A. Synergistic

B. Tendon

C. Latissimus

D. Gastrocnemius

E. Radial

F. Gluteus

3 The full name of any muscle should be used in a clinical setting for several reasons. Even though personnel in a particular profession will be familiar with both the "layman's" and technical terms for muscles, it is important that the full terminology generally be used. For example, there are multiple muscles with the same initial term. If a medical transcriptionist is told that a patient had damage to the biceps, he or she might mistakenly record B. brachii rather than B. femoris. Similarly, if directions for an injection are indicated on the order sheet as just the gluteus muscle, the wrong gluteus might be injected, risking nerve damage. However, when speaking with a patient, it is important to know both the technical terms and layman's terms.

VII. FILL IN THE CHART

MUSCLE	ORIGIN	INSERTION
Flexors of the Elbow		
Biceps brachii	Supraglenoid tubercle and coracoid process of scapula	Radial tuberosity and aponeurosis of flexor muscles of lower arm
Brachialis	Distal half of anterior surface of humerus	Coronoid process and tuberosity of ulna
Brachioradialis	Lateral supracondylar ridge of humerus	Lateral side of styloid process of radius
Pronator teres	Above medial condyle of humerus and medial side of coronoid process of ulna	Middle of lateral surface of radius
Extensors of the Elbow		
Triceps brachii	Infraglenoid tubercle of scapula and medial and posterior surface of humerus	Posterior surface of olecranon process of ulna
Anconeus	Posterior surface of lateral epicondyle of humerus	Lateral surface of olecranon process and posterior proximal surface of ulna

MUSCLE	ORIGIN	INSERTION
Supinators of the Forearm		
Supinator	Lateral epicondyle of humerus and annular and collateral radial ligament	Lateral surface of upper one-third of body of the radius
Biceps brachii	Supraglenoid tubercle and coracoid process of scapula	Radial tuberosity and aponeurosis of flexor muscles of lower arm
Brachioradialis	Lateral supracondylar ridge of humerus	Lateral side of styloid process of radius
Pronators of the Forearm		
Pronator teres	Above the medial epicondyle of humerus and medial side of coronoid process of ulna	Middle of lateral surface of radius
Pronator quadratus	Medial side of anterior surface of the distal one-fourth of ulna	Lateral side of distal one-fourth of the radius
Brachioradialis	Lateral supracondylar ridge of humerus	Lateral side of styloid process of radius

MUSCLE	ORIGIN	INSERTION
Flexors of the Knee		
Gracilis	Inferior ramus and body of pubis	Medial surface of tibia just below the condyle
Semitendinosus	Ischial tuberosity	Upper medial surface of tibia
Semimembranosus	Ischial tuberosity	Posterior part of the medial condyle of tibia
Gastrocnemius	Upper posterior part of medial condyle of tibia	Calcaneus
Sartorius	Anterior superior iliac spine and upper half of iliac notch	Proximal part of the medial aspect of tibia
Biceps femoris	Ischial tuberosity and linea aspera and proximal two-thirds of supracondylar line	Head of fibula and lateral condyle of tibia
Plantaris	Lateral supracondylar ridge of femur	Posterior surface of calcaneus

MUSCLE	ORIGIN	INSERTION
Extensors of the Knee		
Tensor fascia latae	Iliac crest and anterior superior iliac spine	Iliotibial band inserts on the lateral epicondyle of the tibia
Vastus lateralis	Proximal intertrochanteric line, greater trochanter and linea aspera	Patella and tibial tuberosity
Vastus medialis	Lower half of intertrochanteric line, linea aspera, medial supracondylar line	Patella and tibial tuberosity
Vastus intermedius	Anterior and lateral surfaces of proximal two-thirds of femur	Patella and tibial tuberosity
Rectus femoris	Anterior inferior iliac spine and upper margin of acetabulum	Patella and tibial tuberosity

MUSCLE	ORIGIN	INSERTION
Plantar Flexors of the Ankle		
Plantaris	Lateral supracondylar ridge of femur	Posterior surface of calcaneus
Gastrocnemius	Upper posterior part of medial condyle and supracondylar ridge of femur	Calcaneus
Soleus	Upper one-fourth of posterior surface of fibula and tibia	Calcaneus
Fibularis longus	Upper two-thirds of lateral surface of fibula	Lateral side of medial cuneiform and base of first metatarsal
Fibularis brevis	Lower two-thirds of lateral surface of the fibula	Lateral side of the base of the fifth metatarsal
Flexor digitorum longus	Medial part of posterior surface of tibia	Plantar surface of bases of distal phalanges 2–5

MUSCLE	ORIGIN	INSERTION
Dorsiflexors of the Ankle		
Extensor digitorum longus	Lateral condyle of tibia, proximal three-fourths of fibula and interosseous membrane	Base of middle and distal phalanges of digits 2–5
Fibularis tertius	Lower one-third of anterior surface of fibula and interosseous membrane	Dorsal surface of the base of the fifth metatarsal
Tibialis anterior	Lateral condyle of proximal one-half of the lateral surface of tibia and interosseous membrane	Medial plantar surface of first cuneiform and base of first metatarsal
Extensor hallucis longus	Middle half of anterior surface of fibula and interosseous membrane	Base of the distal phalanx of big toe

VIII. CASE STUDIES

A

1. Hamstring muscles; flexion
2. Quadriceps muscles; extension
3. Tibia
4. Tibial tuberosity
5. Osgood-Schlatter disease
6. The tibial tuberosity is underlain by cartilage in the young

B

1. Acetabulum
2. The powerful adductors would pull on the femur
3. Medial rotators of hip, adductors of the hip, and flexors of the knee
4. Hamstring muscles, sartorius, and plantaris
5. Gluteus medius and minimus and tensor fascia latae
6. Sciatic

CHAPTER 7

Nerve Innervation

VI. SHORT ANSWER

1. Cervical nerves 3 and 4 (C3, C4) are part of the cervical plexus. The phrenic nerve arises from fibers from C3, C4, and C5. Thus, if the phrenic nerve were damaged, the diaphragm would not contract, and inhalation would be compromised.
2. The axillary nerve innervates the deltoid and teres minor muscles. Damage to it would impede shoulder movements initiated by these two muscles, including flexion, medial rotation, and abduction of the arm.
3. Compression of the sciatic nerve would produce pain in the buttocks, hamstrings, and lower leg. The patient would have difficulty walking.
4. When the patellar ligament is tapped, stretch receptors initiate a reflex reaction in which sensory impulses to the spinal cord cause motor impulses to be sent to the quadriceps femoris muscle. When it contracts, the lower leg is extended at the knee.

VII. CASE STUDIES

A

1. Bell's palsy
2. Facial (VII)
3. Mandibular, zygomatic, and buccal
4. Levator palpebrae superioris, zygomaticus major and minor, buccinator,

B

1. "Driver's thigh"
2. Gluteal muscles
3. Biceps femoris and tensor fascia latae
4. Sciatic
5. Piriformis syndrome

NOTES

NOTES

NOTES

NOTES